# Algebra II FOR BEGINNERS

## The Ultimate Step by Step Guide to Acing Algebra II

By

Reza Nazari

All inquiries should be addressed to:
info@effortlessMath.com
www.EffortlessMath.com


ISBN: 978-1-63719-263-4

Published by: **Effortless Math Education Inc.**

**for Online Math Practice Visit www.EffortlessMath.co**

# *Welcome to*
# Algebra II Prep
## 2023

Thank you for choosing Effortless Math for your Algebra II preparation and congratulations on making the decision to take the Algebra II course! It's a remarkable move you are taking, one that shouldn't be diminished in any capacity.

That's why you need to use every tool possible to ensure you succeed on the final exam with the highest possible score, and this extensive study guide is one such tool.

*Algebra II for Beginners* is designed to be comprehensive and cover all the topics that are typically covered in an Algebra II course. It provides clear explanations and examples of the concepts and includes practice problems and quizzes to test your understanding of the material. The textbook also provides step-by-step solutions to the problems, so you can check your work and understand how to solve similar problems on your own.

Additionally, this book is written in a user-friendly way, making it easy to follow and understand even if you have struggled with math in the past. It also includes a variety of visual aids such as diagrams, graphs, and charts to help you better understand the concepts.

*Algebra II for Beginners* is flexible and can be used to supplement a traditional classroom setting, or as a standalone resource for self-study. With the help of this comprehensive textbook, you will have the necessary foundation to master the material and succeed in the Algebra II course.

## Effortless Math's Algebra II Online Center

Effortless Math Online Algebra II Center offers a complete study program, including the following:

✓ Step-by-step instructions on how to prepare for the Algebra II test

✓ Numerous Algebra II worksheets to help you measure your math skills

✓ Complete list of Algebra II formulas

✓ Video lessons for all Algebra II topics

✓ Full-length Algebra II practice tests

✓ And much more...

**No Registration Required.**

Visit **EffortlessMath.com/Algebra2** to find your online Algebra II resources.

# How to Use This Book Effectively

Look no further when you need a study guide to improve your math skills to succeed on the Algebra II course. Each chapter of this comprehensive guide to the Algebra II will provide you with the knowledge, tools, and understanding needed for every topic covered on the course.

It's very important that you understand each topic before moving onto another one, as that's the way to guarantee your success. Each chapter provides you with examples and a step-by-step guide of every concept to better understand the content that will be on the course. To get the best possible results from this book:

➢ **Begin studying long before your final exam date**. This provides you ample time to learn the different math concepts. The earlier you begin studying for the test, the sharper your skills will be. Do not procrastinate! Provide yourself with plenty of time to learn the concepts and feel comfortable that you understand them when your test date arrives.

➢ **Practice consistently**. Study Algebra II concepts at least 30 to 40 minutes a day. Remember, slow and steady wins the race, which can be applied to preparing for the Algebra II test. Instead of cramming to tackle everything at once, be patient and learn the math topics in short bursts.

➢ Whenever you get a math problem wrong, **mark it off, and review it later** to make sure you understand the concept.

➢ Start each session by **looking over the previous material.**

➢ Once you've reviewed the book's lessons, **take a practice test at the back of the book** to gauge your level of readiness. Then, review your results. Read detailed answers and solutions for each question you missed.

➢ **Take another practice test** to get an idea of how ready you are to take the actual exam. Taking the practice tests will give you the confidence you need on test day. Simulate the Algebra II testing environment by sitting in a quiet room free from distraction. Make sure to clock yourself with a timer.

# Looking for more?

Visit [EffortlessMath.com/Algebra2](EffortlessMath.com/Algebra2) to find hundreds of Algebra II worksheets, video tutorials, practice tests, Algebra II formulas, and much more.

Or scan this QR code.

**No Registration Required.**

# Contents

## Chapter: **Probability**      **257**

**18**

# 1 Fundamentals and Building Blocks

Math topics that you'll learn in this chapter:

- ☑ Order of Operations
- ☑ Scientific Notation
- ☑ Rules of Exponents
- ☑ Evaluating Expressions
- ☑ Simplifying Algebraic Expressions
- ☑ Sets

1

# Order of Operations

- In Mathematics, "operations" are addition, subtraction, multiplication, division, exponentiation (written as $b^n$), and grouping.

- When there is more than one math operation in an expression, use PEMDAS: (to memorize this rule, remember the phrase "Please Excuse My Dear Aunt Sally".)
  - ❖ Parentheses
  - ❖ Exponents
  - ❖ Multiplication and Division (from left to right)
  - ❖ Addition and Subtraction (from left to right)

## Examples:

**Example 1.** Calculate. $(2 + 6) \div (2^2 \div 4) =$

*Solution*: First, simplify inside parentheses:
$(8) \div (4 \div 4) = (8) \div (1)$.
Then: $(8) \div (1) = 8$.

**Example 2.** Solve. $(6 \times 5) - (14 - 5) =$

*Solution*: First, calculate within parentheses:
$(6 \times 5) - (14 - 5) = (30) - (9)$.
Then: $(30) - (9) = 21$.

**Example 3.** Calculate. $-4[(3 \times 6) \div (3^2 \times 2)] =$

*Solution*: First, calculate within parentheses:
$-4[(18) \div (9 \times 2)] = -4[(18) \div (18)] = -4[1]$.
Multiply $-4$ and $1$, then: $-4[1] = -4$.

**Example 4.** Solve. $(28 \div 7) + (-19 + 3) =$

*Solution*: First, calculate within parentheses:
$(28 \div 7) + (-19 + 3) = (4) + (-16)$, then: $(4) - (16) = -12$.

# Scientific Notation

- Scientific notation is used to write very big or very small numbers in decimal form.

- In scientific notation, all numbers are written in the form of: $m \times 10^n$, where $m$ is greater than 1 and less than 10.

- To convert a number from scientific notation to standard form, move the decimal point to the left (If the exponent of ten is a negative number.), or to the right. (If the exponent is positive.)

## Examples:

**Example 1.** Write 0.00024 in scientific notation.

*Solution*: First, move the decimal point to the right so you have a number between 1 and 10. That number is 2.4. Now, determine how many places the decimal moved in step 1 by the power of 10. We moved the decimal point 4 digits to the right. Then: $10^{-4}$. When the decimal moved to the right, the exponent is negative. Then: $0.00024 = 2.4 \times 10^{-4}$.

**Example 2.** Write $3.8 \times 10^{-5}$ in standard notation.

*Solution*: $10^{-5} \rightarrow$ When the decimal moved to the left, the exponent is negative. Then: $3.8 \times 10^{-5} = 0.000038$.

**Example 3.** Write 0.00031 in scientific notation.

*Solution*: First, move the decimal point to the right so you have a number between 1 and 10. Then: $m = 3.1$. Now, determine how many places the decimal moved in step 1 by the power of 10.
$10^{-4} \rightarrow$ Then: $0.00031 = 3.1 \times 10^{-4}$.

**Example 4.** Write $6.2 \times 10^5$ in standard notation.

*Solution*: $10^5 \rightarrow$ The exponent is positive 5. Then, move the decimal point to the right five digits. (Remember $6.2 = 6.20000$).
Then: $6.2 \times 10^5 = 620,000$.

bit.ly/3VKMHn8

Find more at

# Rules of Exponents

- Exponents are shorthand for repeated multiplication of the same number by itself. For example, instead of $2 \times 2$, we can write $2^2$. For $3 \times 3 \times 3 \times 3$, we can write $3^4$.

- In algebra, a variable is a letter used to stand for a number. The most common letters are: $x$, $y$, $z$, $a$, $b$, $c$, $m$, and $n$.

- Exponent's rules: $x^a \times x^b = x^{a+b}$, $\frac{x^a}{x^b} = x^{a-b}$.

$$(x^a)^b = x^{a \times b} \qquad (xy)^a = x^a \times y^a \qquad \left(\frac{a}{b}\right)^c = \frac{a^c}{b^c}$$

## Examples:

**Example 1.** Multiply. $2x^2 \times 3x^4$

*Solution*: Use Exponent's rules: $x^a \times x^b = x^{a+b} \rightarrow x^2 \times x^4 = x^{2+4} = x^6$.
Then: $2x^2 \times 3x^4 = 6x^6$.

**Example 2.** Simplify. $(x^4y^2)^2$

*Solution*: Use Exponent's rules: $(x^a)^b = x^{a \times b}$.
Then: $(x^4y^2)^2 = x^{4 \times 2}y^{2 \times 2} = x^8y^4$.

**Example 3.** Multiply. $5x^8 \times 6x^5$

*Solution*: Use Exponent's rules: $x^a \times x^b = x^{a+b} \rightarrow x^8 \times x^5 = x^{8+5} = x^{13}$.
Then: $5x^8 \times 6x^5 = 30x^{13}$.

**Example 4.** Simplify. $(x^4y^7)^3$

*Solution*: Use Exponent's rules: $(x^a)^b = x^{a \times b}$.
Then: $(x^4y^7)^3 = x^{4 \times 3}y^{7 \times 3} = x^{12}y^{21}$.

**Example 5.** Simplify. $\frac{xy^2}{x^3y}$

*Solution*: Use Exponent's rules: $\frac{x^a}{x^b} = x^{a-b} \rightarrow \frac{x}{x^3} = x^{-2}$ and $\frac{y^2}{y} = y$.
Then: $\frac{xy^2}{x^3y} = \frac{y}{x^2} = yx^{-2}$.

## Evaluating Expressions

- To evaluate an algebraic expression, substitute a number for each variable.

- Perform the arithmetic operations to find the value of the expression.

## Examples:

**Example 1.** Calculate this expression for $a = 2$ and $b = -1$. $4a - 3b$

*Solution*: First, substitute 2 for $a$, and $-1$ for $b$.
Then:
$4a - 3b = 4(2) - 3(-1)$.
Now, use order of operation to find the answer:
$4(2) - 3(-1) = 8 + 3 = 11$.

**Example 2.** Evaluate this expression for $x = -2$ and $y = 2$. $3x + 6y$

*Solution*: Substitute $-2$ for $x$, and 2 for $y$.
Then:
$3x + 6y = 3(-2) + 6(2) = -6 + 12 = 6$.

**Example 3.** Find the value of this expression $2(6a - 5b)$ when $a = -1$ and $b = 4$.

*Solution*: Substitute $-1$ for $a$, and 4 for $b$.
Then:
$2(6a - 5b) = 12a - 10b = 12(-1) - 10(4) = -12 - 40 = -52$.

**Example 4.** Solve this expression. $-7x - 2y$ when $x = 4$ and $y = -3$.

*Solution*: Substitute 4 for $x$, and $-3$ for $y$ and simplify.
Then:
$-7x - 2y = -7(4) - 2(-3) = -28 + 6 = -22$.

**Example 5.** Find the value of $4ab + b^2$, when $a = 3$ and $b = -1$.

*Solution*: Substitute 3 for $a$ and $-1$ for $b$. Then:

$4(3)(-1) + (-1)^2 = -12 + 1 = -11$.

bit.ly/3H5ijOP

Find more at

# Simplifying Algebraic Expressions

- In algebra, a variable is a letter used to stand for a number. The most common letters are $x$, $y$, $z$, $a$, $b$, $c$, $m$, and $n$.

- An algebraic expression is an expression that contains integers, variables, and math operations such as addition, subtraction, multiplication, division, etc.

- In an expression, we can combine "like" terms. (Values with same variable and same power)

## Examples:

**Example 1.** Simplify. $(4x + 2x + 4) =$

*Solution*: In this expression, there are three terms: $4x$, $2x$, and $4$. Two terms are "like" terms: $4x$ and $2x$. Combine like terms: $4x + 2x = 6x$.
Then: $(4x + 2x + 4) = 6x + 4$.
(Remember you cannot combine variables and numbers.)

**Example 2.** Simplify. $-2x^2 - 5x + 4x^2 - 9 =$

*Solution*: Combine "like" terms: $-2x^2 + 4x^2 = 2x^2$.

Then: $-2x^2 - 5x + 4x^2 - 9 = 2x^2 - 5x - 9$.

**Example 3.** Simplify. $(-8 + 6x^2 + 3x^2 + 9x) =$

*Solution*: Combine like terms. Then:
$(-8 + 6x^2 + 3x^2 + 9x) = 9x^2 + 9x - 8$.

**Example 4.** Simplify. $-10x + 6x^2 - 3x + 9x^2 =$

*Solution*: Combine "like" terms: $-10x - 3x = -13x$, and $6x^2 + 9x^2 = 15x^2$.

Then: $-10x + 6x^2 - 3x + 9x^2 = -13x + 15x^2$. Write in standard form (Biggest powers first): $-13x + 15x^2 = 15x^2 - 13x$.

# Sets

- In mathematics, a "Set" is a collection or a group of distinct members (separate from each other) that have a common feature and this feature can be easily described and defined. The sets show capital English letters like $A$, $B$, and etc. The members or elements of the set are placed inside the braces { }.

- We show the sign of membership or being a member of a set with the symbol $\in$ and the sign of not being a member of a set with $\notin$.

- The members of a set must be non-repeating, so in the set, repeating members are counted only once. In the set, the order of the members is not important, that is, by moving the members of a set, a new set is not defined.

- If the number of members of a set is countable, we say that set is finite, and if the number of members of a set is uncountable, we say that set is infinite.

  An empty or null set is a set that has no members. The null set is represented by the symbol $\emptyset$ or { }.

## Examples:

**Example 1.** Find the elements of the set of $A$ represent the first 5 multiples of 6.

**Solution**: The question asks us to find the first 5 multiples of 6, so we find the members of the set using the multiplication operation: $6 \times 1 = 6$, $6 \times 2 = 12$, $6 \times 3 = 18$, $6 \times 4 = 24$, $6 \times 5 = 30$. So, we have set $A = \{6,12,18,24,30\}$.

**Example 2.** Find the elements of the set of $B$ that represent the odd numbers between 12 and 20.

**Solution**: The question asks us to find the odd numbers between 12 and 20, so we easily list the odd numbers between these two numbers: 13, 15, 17, 19. So, we have set $B = \{13,15,17,19\}$.

**Example 3.** Find the elements of the set of $C$ that represent the even whole less than 9.

**Solution**: The even whole numbers less than 9 begin from zero and end at 8. So, we have set $C = \{0,2,4,6,8\}$.

bit.ly/3QjXsM6

Find more at

# Chapter 1: Practices

✎ **Evaluate each expression.**

1) $12 + (3 \times 2) =$ 14

2) $8 - (4 \times 5) =$ −12

3) $(8 \times 2) + 14 =$ 32

4) $(10 - 6) - (4 \times 3) =$ −8

5) $15 + (12 \div 2) =$ 21

6) $(24 \times 3) \div 4 =$ 18

7) $(28 \div 2) \times (-4) =$ −56

8) $(2 \times 6) + (14 - 8) =$ 18

9) $45 + (4 \times 2) + 12 =$ 65

10) $(10 \times 5) \div (4 + 1) =$ 10

11) $(-6) + (8 \times 6) + 10 =$ 52

12) $(12 \times 4) - (56 \div 4) =$ 34

✎ **Write each number in scientific notation.**

13) $15,000,000 =$ $1.5 \times 10^7$

14) $67,000 =$ $6.7 \times 10^4$

15) $0.000819 =$ $8.19 \times 10^{-4}$

16) $0.00092 =$ $9.2 \times 10^{-4}$

I don't think I did this right...

✎ **Write each number in standard notation.**

17) $4.5 \times 10^3 =$ 4500

18) $8 \times 10^{-4} =$ .0008?

19) $6 \times 10^{-1} =$ .6

20) $9 \times 10^{-2} =$ .09

✎ **Simplify and write the answer in exponential form.**

21) $2x^2 \times 4x =$ $4x^2 \times 4x$

22) $5x^4 \times x^2 =$ $5x^6$

23) $8x^4 \times 3x^5 =$ $24x^9$

24) $3x^2 \times 6xy =$ $18x^3 y$

25) $2x^5y \times 4x^2y^3 =$ $8x^4 y^4$

26) $9x^2y^5 \times 5x^2y^8 =$ $45x^4 y^{13}$

27) $5x^2y \times 5x^2y^7 =$ $25x^4 y^8$

28) $7x^6 \times 3x^9y^4 =$ $21x^{15} y^4$

29) $8x^8y^5 \times 7x^5y^3 =$ $56x^{13} y^4$

30) $9x^6x^2 \times 4xy^5 =$ $36x^9 y^5$

31) $12xy^7 \times 2x^9y^8 =$ $24x^{10} y^{15}$

32) $9x^9y^{12} \times 9x^{14}y^{11} =$ $81x^{23} y^{23}$

haven't done this in a year & a half either...

Effortless Math Education

**Evaluate each expression using the values given.**

33) $x + 2y$, $x = 1$, $y = 2$ *5*

34) $2x - 3y$, $x = 1$, $y = -2$ *-4*

35) $-a + 5b$, $a = -2$, $b = 3$ *13*

36) $-3a + 5b$, $a = 5$, $b = 2$ *-5*

37) $5x + 8 - 3y$, $x = 5$, $y = 4$ *21*

38) $3x + 5y$, $x = 2$, $y = 3$ *21*

39) $7x + 6y$, $x = 2$, $y = 4$ *34*

40) $3a - (12 - b)$, $a = 3$, $b = 5$ *2*

41) $4z + 20 + 7k$, $z = -4$, $k = 5$ *39*

42) $xy + 15 + 4x$, $x = 6$, $y = 3$ *54*

43) $8x + 3 - 5y + 4$, $x = 6$, $y = 3$ *40*

44) $5 + 2(-3x - 4y)$, $x = 6$, $y = 5$
$7 (-3(6) - 4(5))$
$7(-18 - (20))$
$7(2)$
$\boxed{14}$

**Simplify each expression.**

45) $x - 4 + 6 - 2x =$ *$-x + 2$*

46) $3 - 4x + 14 - 3x =$ *$14 - 7x$*

47) $33x - 5 + 13 + 4x =$ *$37x + 8$*

48) $-3 - x^2 - 7x^2 =$ *$-8x^2 - 3$*

49) $4 + 11x^2 + 3 =$ *$11x^2 + 4$*

50) $7x^2 + 5x + 6x^2 =$ *$13x^2 + 5x$*

51) $42x + 15 + 3x^2 =$ *$3x^2 + 42x + 15$*

52) $6x(x - 2) - 5 =$ *$6x^2 - 12x - 5$*

53) $7x - 6 + 9x + 3x^2 =$ *$3x^2 + 16x - 6$*

54) $(-5)(7x - 2) + 12x =$ *$-23x + 10$*
*$15x - 36 + 42x$*

55) $15x - 6(6 - 7x) =$ *$57x - 36$*
*$25x + 42x + 30 - 4$*

56) $25x + 6(7x + 5) - 4 =$ *$67x + 26$*

**Write the following sets in the roster form.** *Never done this before...*

*Ik this is what you wanted....?*

57) The set of all even numbers less than 14. *$-14, -12, -10, -8, -6, -4, -2$* $\boxed{0, 2, 4, 6, 10, 12}$

58) The set of first 5 odd numbers. *$1, 3, 5, 7, 9$?*

59) The set of all factors of 24. *$1, 2, 3, 4, 6, 8, 12, 24$*

60) The set of all factors of 36. *$1, 2, 3, 4, 6, 8, 9, 12, 18, 36$*

*$-3 - x^2 - 4x^2$*

Effortless Math Education

# Chapter 1: Answers

1) 18

2) $-12$

3) 30

4) $-8$

5) 21

6) 18

7) $-56$

8) 18

9) 65

10) 10

11) 52

12) 34

13) $1.5 \times 10^7$

14) $6.7 \times 10^4$

15) $8.19 \times 10^{-4}$

16) $9.2 \times 10^{-4}$

17) 4,500

18) 0.0008

19) 0.6

20) 0.09

21) $8x^3$

22) $5x^6$

23) $24x^9$

24) $18x^3y$

25) $8x^7y^4$

26) $45x^4y^{13}$

27) $25x^4y^8$

28) $21x^{15}y^4$

29) $56x^{13}y^8$

30) $36x^9y^5$

31) $24x^{10}y^{15}$

32) $81x^{23}y^{23}$

33) 5

34) 8

35) 17

36) $-5$

37) 21

38) 21

39) 38

40) 2

41) 39

42) 57

43) 40

44) $-71$

45) $-x + 2$

46) $-7x + 17$

47) $37x + 8$

48) $-8x^2 - 3$

49) $11x^2 + 7$

50) $13x^2 + 5x$

51) $3x^2 + 42x + 15$

52) $6x^2 - 12x - 5$

53) $3x^2 + 16x - 6$

54) $-23x + 10$

55) $57x - 36$

56) $67x + 26$

57) $\{2, 4, 6, 8, 10, 12\}$

58) $\{1, 3, 5, 7, 9\}$

59) $\{1, 2, 3, 4, 6, 8, 12, 24\}$

60) $\{1, 2, 3, 4, 6, 9, 12, 18, 36\}$

**Effortless Math Education**

# CHAPTER

# 2 Equations and Inequalities

Math topics that you'll learn in this chapter:

- ☑ Solving Multi– Step Equations
- ☑ Slope and Intercepts
- ☑ Using Intercepts
- ☑ Transforming Linear Functions
- ☑ Solving Inequalities
- ☑ Graphing Linear Inequalities
- ☑ Solving Compound Inequalities
- ☑ Solving Absolute Value Equations
- ☑ Solving Absolute Value Inequalities
- ☑ Graphing Absolute Value Inequalities
- ☑ Solving Systems of Equations
- ☑ Solving Special Systems
- ☑ Systems of Equations Word Problems

11

# Solving Multi–Step Equations

- To solve a multi-step equation, combine "like" terms on one side.

- Bring variables to one side by adding or subtracting.

- Simplify using the inverse of addition or subtraction.

- Simplify further by using the inverse of multiplication or division.

- Check your solution by plugging the value of the variable into the original equation.

## Examples:

**Example 1.** Solve this equation for $x$. $4x + 8 = 20 - 2x$

*Solution:* First, bring variables to one side by adding $2x$ to both sides. Then:

$4x + 8 + 2x = 20 - 2x + 2x \rightarrow 4x + 8 + 2x = 20$.

Simplify: $6x + 8 = 20$.

Now, subtract 8 from both sides of the equation:

$6x + 8 - 8 = 20 - 8 \rightarrow 6x = 12$.

Divide both sides by 6:

$6x = 12 \rightarrow \frac{6x}{6} = \frac{12}{6} \rightarrow x = 2$.

Let's check this solution by substituting the value of 2 for $x$ in the original equation:

$x = 2 \rightarrow 4x + 8 = 20 - 2x \rightarrow 4(2) + 8 = 20 - 2(2) \rightarrow 16 = 16$.

The answer $x = 2$ is correct.

**Example 2.** Solve this equation for $x$. $-5x + 4 = 24$

*Solution:* Subtract 4 from both sides of the equation.

$-5x + 4 = 24 \rightarrow -5x + 4 - 4 = 24 - 4 \rightarrow -5x = 20$.

Divide both sides by $-5$, then:

$-5x = 20 \rightarrow \frac{-5x}{-5} = \frac{20}{-5} \rightarrow x = -4$.

Now, check the solution:

$x = -4 \rightarrow -5x + 4 = 24 \rightarrow -5(-4) + 4 = 24 \rightarrow 24 = 24$.

The answer $x = -4$ is correct.

bit.ly/3nQbSEB

Find more at

# Slope and Intercepts

- The slope of a line represents the direction of a line on the coordinate plane.

- A coordinate plane contains two perpendicular number lines. The horizontal line is $x$ and the vertical line is $y$. The point at which the two axes intersect is called the origin. An ordered pair $(x, y)$ shows the location of a point.

- A line on a coordinate plane can be drawn by connecting two points.

- To find the slope of a line, we need the equation of the line or two points on the line.

- The slope of a line with two points $A(x_1, y_1)$ and $B(x_2, y_2)$ can be found by using this formula: $\frac{y_2-y_1}{x_2-x_1} = \frac{rise}{run}$.

- The equation of a line is typically written as $y = mx + b$ where $m$ is the slope and $b$ is the $y$−intercept.

## Examples:

**Example 1.** Find the slope of the line through these two points: $A(1, -6)$ and $B(3,2)$.

**Solution**: Slope $= \frac{y_2-y_1}{x_2-x_1}$. Let $(x_1, y_1)$ be $A(1, -6)$ and $(x_2, y_2)$ be $B(3,2)$.

(Remember, you can choose any point for $(x_1, y_1)$ and $(x_2, y_2)$.)

Then:

Slope $= \frac{y_2-y_1}{x_2-x_1} = \frac{2-(-6)}{3-1} = \frac{8}{2} = 4$.

The slope of the line through these two points is 4.

**Example 2.** Find the slope of the line with equation $y = -2x + 8$.

**Solution**: when the equation of a line is written in the form of $y = mx + b$, the slope is $m$.

In this line: $y = -2x + 8$, the slope is $-2$.

# Using Intercepts

- The cartesian plane is the $2-$dimensional coordinate plane that consists of a horizontal number line that is called the $x-$axis, and a perpendicular number line through the point 0 on the first number line which is called the $y-$axis.

- Intercepts are points where a graph passes through an axis. An $x-$intercept is where the graph passes through the $x-$axis. A $y-$intercept is where the graph passes through the $y-$axis.

- The $y-$axis is the line where $x$ is equal to zero and the $x-$axis is the line where $y$ is equal to zero. An $x-$intercept can be considered as a point on the graph where the value of $y$ is zero, and a $y-$intercept can be considered as a point on the graph where the $x-$value is zero.

- When the equation is represented in the slope$-$intercept form $y = mx + b$, the $y-$intercept is equal to the value of $b$. When you consider the $x = 0$, then $mx$ becomes equal to 0, so when the value of $x$ is equal to 0, $y = b$.

- When you want to find the $x-$intercept, you can put the value of $y$ equal to 0 and solve the equation for $x$. In this way, when the value of $y$ is equal to 0, the line passes through the $x-$axis.

- When the given equation isn't in $y = mx + b$ form, you can find the value of the intercepts by putting in 0 where needed and solving for the other variable.

## Examples:

**Example 1.** Find the $x-$intercept and $y-$intercept of the following line:
$$5x + 4y = 40$$
***Solution***: Find the $x-$intercept by replacing $y$ with 0 and solve for $x$: $5x + 4y = 40 \rightarrow 5x + 4(0) = 40 \rightarrow 5x + 0 = 40 \rightarrow 5x = 40 \rightarrow x = \frac{40}{5} = 8 \rightarrow x = 8$.

So, the $x-$intercept is 8. Find the $y-$intercept by replacing $x$ with 0 and solve for $y$: $5x + 4y = 40 \rightarrow 5(0) + 4y = 40 \rightarrow 0 + 4y = 40 \rightarrow y = \frac{40}{4} = 10$.

So, the $y-$intercept is 10.

**Example 2.** Find the $x-$intercept and $y-$intercept of the following line:
$$8x^2 + 2y^2 = 32.$$
***Solution***: Find the $x-$intercept by replacing $y$ with 0 and solve for $x$: $8x^2 + 2y^2 = 32 \rightarrow 8x^2 + 2(0)^2 = 32 \rightarrow 8x^2 + 0 = 32 \rightarrow 8x^2 = 32 \rightarrow x^2 = \frac{32}{8} = 4 \rightarrow x = \pm2$.

So, the $x-$intercept is $\pm2$. Then find the $y-$intercept by replacing $x$ with 0 and solve for $y$: $8x^2 + 2y^2 = 32 \rightarrow 8(0)^2 + 2y^2 = 32 \rightarrow 0 + 2y^2 = 32 \rightarrow y^2 = \frac{32}{2} = 16 \rightarrow y = \pm4$. So, the $y-$intercept is $\pm4$.

# Transforming Linear Functions

- Transformation of linear functions means moving the graphs of linear functions by changing the position of the $y$-intercept or the line's slope without changing the line's shape. The transformed linear function still follows the slope-intercept form $(y = mx + b)$. The three types of linear function transformations are translation, rotation, and reflection. In fact, a transformation of the linear function is a change in the position or size of a linear function.

- A group of functions whose graphs have basic common features is called a family of functions.

- The most basic function in a family of functions is called the parent function. The parent function of linear functions is $f(x) = x$.

- All the graphs of linear functions are transformed shapes of the parent function's graph $(f(x) = x)$.

- The graph is translated vertically when you change the $y$-intercept $(b)$ in the function $f(x) = mx + b$. If the value of $b$ rises up, the graph is translated up. If the value of $b$ reduces, the graph is translated down.

- The rotation of the graph about the point $(0, b)$ occurs when you change the slope $m$ in the function $f(x) = mx + b$. In this case, the rotation of the graph changes the steepness of the line.

- The reflection of the graph across the $y$-axis occurs when you multiply the slope $m$ of function $f(x) = mx + b$ by $-1$.

## Example:

Graph $f(x) = x$ and find $g(x) = x + 2$.

**Solution**: Here, the parent function is $f(x) = x$ which passes through the origin $(0,0)$. You should change $y$-intercept $(b)$ in the function. $f(x) = mx + b$:

$f(x) = x + 0 \rightarrow f(x) = x + 2$.

In fact, the value of $b$ rises up 2 units, and the graph is translated up.

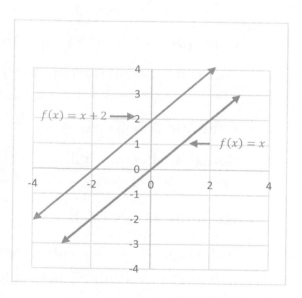

bit.ly/3ZZmyUM

Find more at

# Solving Inequalities

- An inequality compares two expressions using an inequality sign.

- Inequality signs are: "less than" <, "greater than" >, "less than or equal to" ≤, and "greater than or equal to" ≥.

- You only need to perform one Math operation to solve the one−step inequalities.

- To solve one−step inequalities, find the inverse (opposite) operation is being performed.

- For dividing or multiplying both sides by negative numbers, flip the direction of the inequality sign.

## Examples:

**Example 1.** Solve this inequality for $x$. $x + 5 \geq 4$

*Solution*: The inverse (opposite) operation of addition is subtraction. In this inequality, 5 is added to $x$. To isolate $x$ we need to subtract 5 from both sides of the inequality. Then: $x + 5 \geq 4 \rightarrow x + 5 - 5 \geq 4 - 5 \rightarrow x \geq -1$.
The solution is: $x \geq -1$.

**Example 2.** Solve the inequality. $x - 3 > -6$

*Solution*: 3 is subtracted from $x$. Add 3 to both sides.
$x - 3 > -6 \rightarrow x - 3 + 3 > -6 + 3 \rightarrow x > -3$.

**Example 3.** Solve. $4x \leq -8$

*Solution*: 4 is multiplied by $x$. Divide both sides by 4.
Then: $4x \leq -8 \rightarrow \frac{4x}{4} \leq \frac{-8}{4} \rightarrow x \leq -2$.

**Example 4.** Solve. $-3x \leq 6$

*Solution*: $-3$ is multiplied by $x$. Divide both sides by $-3$. Remember when dividing or multiplying both sides of an inequality by negative numbers, flip the direction of the inequality sign.
Then: $-3x \leq 6 \rightarrow \frac{-3x}{-3} \geq \frac{6}{-3} \rightarrow x \geq -2$.

# Graphing Linear Inequalities

- To graph a linear inequality, first draw a graph of the "equals" line.

- Use a dashed line for "less than" ($<$) and "greater than" ($>$) signs and a solid line for "less than and equal to" ($\leq$) and "greater than and equal to" ($\geq$).

- Choose a testing point. (It can be any point on both sides of the line.)

- Put the value of $(x, y)$ of that point in the inequality. If that works, that part of the line is the solution. If the values don't work, then the other part of the line is the solution.

## Example:

Sketch the graph of inequality: $y < 2x + 4$.

**Solution**: To draw the graph of $y < 2x + 4$, you first need to graph the line:

$y = 2x + 4$.

Since there is a "less than" ($<$) sign, draw a dash line.

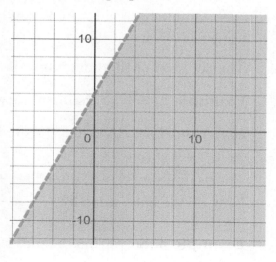

The slope is 2 and the $y$ −intercept is 4.

Then, choose a testing point and substitute the value of $x$ and $y$ from that point into the inequality.

The easiest point to test is the origin, $(0, 0)$:

$(0, 0) \rightarrow y < 2x + 4 \rightarrow 0 < 2(0) + 4 \rightarrow 0 < 4$.

This is correct! 0 is less than 4.

So, this part of the line (on the right side) is the solution of this inequality.

bit.ly/3vZZDet

Find more at

# Solving Compound Inequalities

- To solve a multi–step inequality, combine "like" terms on one side.

- Bring variables to one side by adding or subtracting.

- Isolate the variable.

- Simplify using the inverse of addition or subtraction.

- Simplify further by using the inverse of multiplication or division.

- For dividing or multiplying both sides by negative numbers, flip the direction of the inequality sign.

## Examples:

**Example 1.** Solve this inequality. $8x - 2 \leq 14$

*Solution*: In this inequality, 2 is subtracted from $8x$. The inverse of subtraction is addition. Add 2 to both sides of the inequality:
$8x - 2 + 2 \leq 14 + 2 \rightarrow 8x \leq 16$.
Now, divide both sides by 8. Then:
$8x \leq 16 \rightarrow \frac{8x}{8} \leq \frac{16}{8} \rightarrow x \leq 2$.
The solution of this inequality is $x \leq 2$.

**Example 2.** Solve this inequality. $3x + 9 < 12$

*Solution*: First, subtract 9 from both sides:
$3x + 9 - 9 < 12 - 9$.
Then simplify: $3x + 9 - 9 < 12 - 9 \rightarrow 3x < 3$.
Now divide both sides by 3: $\frac{3x}{3} < \frac{3}{3} \rightarrow x < 1$.

**Example 3.** Solve this inequality. $-5x + 3 \geq 8$

*Solution*: First, subtract 3 from both sides:
$-5x + 3 - 3 \geq 8 - 3 \rightarrow -5x \geq 5$.

Divide both sides by $-5$. Remember that you need to flip the direction of the inequality sign. $-5x \geq 5 \rightarrow \frac{-5x}{-5} \leq \frac{5}{-5} \rightarrow x \leq -1$.

# Solving Absolute Value Equations

- Isolate the absolute value.

- Take off the absolute value sign and solve the equation.

- Write another equation with the negative sign of the answer of the absolute value equation.

- Plug in the value of the variable into the original equation and check your answer.

## Examples:

**Example 1.** Solve this equation. $|x + 1| = 2$

*Solution*: First take off the absolute value sign: $x + 1 = 2$ or $x + 1 = -2$.

Then:

$x = 2 - 1 = 1$ or $x = -2 - 1 = -3$.

**Example 2.** Solve this equation. $|x + 2| = 5$

*Solution*: First take off the absolute value sign $x + 2 = 5$ or $x + 2 = -5$.

Then:

$x = 5 - 2 = 3$ or $x = -5 - 2 = -7$.

**Example 3.** Solve this equation. $|x + 3| = 8$

*Solution*: First take off the absolute value sign $x + 3 = 8$ or $x + 3 = -8$.

Then:

$x = 8 - 3 = 5$ or $x = -8 - 3 = -11$.

**Example 4.** Solve this equation. $|2 - x| = 1$

*Solution*: First take off the absolute value sign $2 - x = 1$ or $2 - x = -1$.

Then:

$x = 2 - 1 = 1$ or $x = 2 + 1 = 3$.

# Solving Absolute Value Inequalities

- An Absolute value can never be negative. Therefore, the absolute value cannot be less than a negative number.

- To solve an absolute value inequality, first isolate it on one side of the inequality. Then, if the inequality sign is greater (or greater and equal), write the value inside the absolute value greater than the value provided and less than the negative of the value provided and solve. If the sign is less than, then do the opposite.

## Examples:

**Example 1.** Solve this equation. $|2x| \geq 24$

*Solution*: Since the inequality sign is greater and equal:

$2x \geq 24$ or $2x \leq -24$.

Then:

$x \geq 12$ or $x \leq -12$.

**Example 2.** Solve this equation. $|x + 2| > 5$

*Solution*: Since the inequality sign is greater than:

$x + 2 > 5$ or $x + 2 < -5$.

Then:

$x > 3$ or $x < -7$.

**Example 3.** Solve this equation. $|1 - x| < 4$

*Solution*: Since the inequality sign is less than:

$$-4 < 1 - x < 4 \rightarrow -5 < -x < 3.$$

Then: $-3 < x < 5$.

# Graphing Absolute Value Inequalities

- Solve the absolute value inequality using the Properties of Inequality.

- Find two key values.

- Represent absolute value inequalities on a number line.

## Examples:

**Example 1.** Solve and graph this equation $|-8x| < 32$.

*Solution*: Use absolute rule: if $|u| < a$, $a > 0$, then: $-a < u < a$.

Therefore: $-32 < -8x < 32$, in other words $-8x > -32$ and $-8x < 32$.

So, we have: $-8x > -32 \rightarrow x < 4$ and $-8x < 32 \rightarrow x > -4$.

Then: $-4 < x < 4$.

**Example 2.** Solve and graph this equation $|10 + 4x| < 14$.

*Solution*: Use absolute rule: if $|u| < a$, $a > 0$, then: $-a < u < a$.

Then: $-14 < 10 + 4x < 14$, $10 + 4x > -14$ and $10 + 4x < 14$.
Thus: $10 + 4x > -14 \rightarrow x > -6$ and $10 + 4x < 14 \rightarrow x < 1$.
Then: $-6 < x < 1$.

**Example 3.** Solve and graph this equation $|2x - 3| \geq 5$.

*Solution*: Use absolute rule: if $|u| \geq a$, $a > 0$, then: $u \geq a$ or $u \leq -a$.

Therefore: $2x - 3 \geq 5$ or $2x - 3 \leq -5$.

So: $x \geq 4$ or $x \leq -1$.

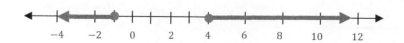

# Solving Systems of Equations

- A system of equations contains two equations and two variables. For example, consider the system of equations: $x - y = 1$, $x + y = 5$.

- The easiest way to solve a system of equations is using the elimination method. The elimination method uses the addition property of equality. You can add the same value to each side of an equation.

- For the first equation above, you can add $x + y$ to the left side and 5 to the right side of the first equation: $x - y + (x + y) = 1 + 5$. Now, if you simplify, you get: $x - y + (x + y) = 1 + 5 \rightarrow 2x = 6 \rightarrow x = 3$. Now, substitute 3 for the $x$ in the first equation: $3 - y = 1$. By solving this equation, $y = 2$.

## Examples:

**Example 1.** What is the value of $x + y$ in this system of equations?

$$\begin{cases} 2x + 4y = 12 \\ 4x - 2y = -16 \end{cases}$$

*Solution*: Solving a system of equations by elimination:
Multiply the first equation by $(-2)$, then add it to the second equation.

$$\begin{array}{c} -2(2x + 4y = 12) \\ 4x - 2y = -16 \end{array} \rightarrow \begin{array}{c} -4x - 8y = -24 \\ 4x - 2y = -16 \end{array} \rightarrow -10y = -40 \rightarrow y = 4.$$

Plug in the value of $y$ into one of the equations and solve for $x$.
$2x + 4(4) = 12 \rightarrow 2x + 16 = 12 \rightarrow 2x = -4 \rightarrow x = -2.$
Thus: $x + y = -2 + 4 = 2.$

**Example 2.** What is the value of $y$ in the following system of equations?

$$\begin{cases} 2x + 5y = 11 \\ 2x - y = -7 \end{cases}$$

*Solution*: Solving systems of equations by elimination: multiply the first equation by $(-1)$, then add it to the second equation.

$$\begin{array}{c} (-1)(2x + 5y = 11) \\ 2x - y = -7 \end{array} \rightarrow \begin{array}{c} -2x - 5y = -11 \\ 2x - y = -7 \end{array} \rightarrow -6y = -18 \rightarrow y = 3.$$

# Solving Special Systems

- 2 linear equations whose graph consists of parallel lines or have an infinite number of solutions may make a special system. For solving the system, you should add or subtract the two linear equations and find the value of the variables $x$ and $y$.

- System of equations with no solution: When the system of equations graph consists of parallel lines or there is no intersection point, the system of the equation has no solution.

- System of equations with infinite number solutions: When the system of equations graph consists of straight lines overlapping each other or there are infinite points as a solution set, the equations system has an infinite number of solutions.

- Three ways for solving special systems of linear equations with two variables:

  - Substitution Method: In this method, you have 2 linear equations in $x$ and $y$; you can represent $y$ in terms of $x$ in one of the equations and then substitute that equation in the second equation.

  - Elimination Method: In this method, you should eliminate one variable. For this purpose, multiply equations by suitable numbers and make one of the variables coefficients the same.

  - Graphical Method: In the graphical method, you can find the solution to the system of equations by plotting their graphs. If your graph consists of parallel lines or straight lines overlapping each other, your system of equations is a special one.

## Example:

Determine if the following system of equations has a solution or not:

$$\left\{ y - 2x = 6, \frac{1}{2}y - x = 8 \right.$$

**Solution**: You can write both equations in the form $y = mx + b$: $\left\{ y - 2x = 6, \right.$ $\frac{1}{2}y - x = 8 \rightarrow y - 2x = 6 \rightarrow y = 2x + 6$ and $\frac{1}{2}y - x = 8 \rightarrow \frac{1}{2}y = x + 8$. Now, use the substitution method and substitute the first equation ($y = 2x + 6$) for $y$ in the second equation ($\frac{1}{2}y - x = 8$): $\frac{1}{2}(2x + 6) - x = 8 \rightarrow x + 3 - x = 8 \rightarrow 3 \neq 8$. This system has no solution so it's a special system.

bit.ly/3XSuvtc

Find more at

# Systems of Equations Word Problems

- Define your variables, write two equations, and use the elimination method for solving systems of equations.

## Examples:

**Example 1.** Tickets to a movie cost $8 for adults and $5 for students. A group of friends purchased 20 tickets for $115.00. How many adults ticket did they buy?

*Solution*: Let $x$ be the number of adult tickets and $y$ be the number of student tickets. There are 20 tickets. Then: $x + y = 20$. The cost of adult tickets is $8 and for students it is $5, and the total cost is $115. So, $8x + 5y = 115$.

Now, we have a system of equations: $\begin{cases} x + y = 20 \\ 8x + 5y = 115 \end{cases}$.

Multiply the first equation by $-5$ and add to the second equation:

$-5(x + y = 20) \rightarrow -5x - 5y = -100$.

$8x + 5y + (-5x - 5y) = 115 - 100 \rightarrow 3x = 15 \rightarrow x = 5 \rightarrow 5 + y = 20 \rightarrow y = 15$. There are 5 adult tickets and 15 student tickets.

**Example 2.** A group of 38 graduates of a high school are going to a picnic in 8 cars. Some of the cars can hold 4 students, and the rest can hold 6 students each. Assuming all the cars are filled, how many of the cars can hold 4 students?

*Solution*: Let $a$ be the number of cars with 4 people capacity and $b$ the number of cars with 6 people capacity. There are 8 cars. Then: $a + b = 8$.

All students are 38, to go with 4 and 6 capacity cars. This means that: $4a + 6b = 38$. Now, we have: $\begin{cases} a + b = 8 \\ 4a + 6b = 38 \end{cases}$.

Solving by elimination method: multiply the first equation by $(-4)$, then add it to the second equation.

$\begin{matrix} (-4)(a + b = 8) \\ 4a + 6b = 38 \end{matrix} \rightarrow \begin{matrix} -4a - 4b = -32 \\ 4a + 6b = 38 \end{matrix} \rightarrow 2b = 6 \rightarrow b = 3$.

Plug in the value of $b$ into one of the equations and solve for $a$.

$a + 3 = 8 \rightarrow a = 5$.

*EffortlessMath.com*

# Chapter 2: Practices

✍ **Solve each equation.**

1) $-3(2 + x) = 3$

2) $-2(4 + x) = 4$

3) $20 = -(x - 8)$

4) $2(2 - 2x) = 20$

5) $-12 = -(2x + 8)$

6) $5(2 + x) = 5$

7) $2(x - 14) = 4$

8) $-28 = 2x + 12x$

9) $3x + 15 = -x - 5$

10) $2(3 + 2x) = -18$

11) $12 - 2x = -8 - x$

12) $10 - 3x = 14 + x$

✍ **Find the slope of the line through each pair of points.**

13) $(1,1), (2,3)$

14) $(-1,2), (0,3)$

15) $(3,-1), (2,3)$

16) $(-2,-1), (0,5)$

17) $(5,1), (2,4)$

18) $(-3,1), (-2,4)$

19) $(6,2), (7,4)$

20) $(6,-5), (3,4)$

21) $(12,-9), (11,-8)$

22) $(7,4), (5,-2)$

23) $(1,1), (3,5)$

24) $(7,-12), (5,10)$

✍ **Solve each inequality and graph it.**

25) $2x \geq 12$

-16   -12   -8   -4   0   4   8   12   16

26) $4 + x \leq 5$

-16   -12   -8   -4   0   4   8   12   16

**Effortless Math Education**

27) $x + 3 \leq -3$

28) $4x \geq 16$

29) $9x \leq 18$

✎ **Find the $x$ −intercepts and $y$ −intercepts of the line.**

30) $2x + 8y = -8$

   $x$ −intercepts: _____

   $y$ −intercepts: _____

31) $3x - 2y = 24$

   $x$ −intercepts: _____

   $y$ −intercepts: _____

32) $-3x + 5y = -15$

   $x$ −intercepts: _____

   $y$ −intercepts: _____

33) $8x - 2y = 10$

   $x$ −intercepts: _____

   $y$ −intercepts: _____

✎ **Graph $f(x)$ and find $g(x)$.**

34) $f(x) = x, g(x) = x - 3$

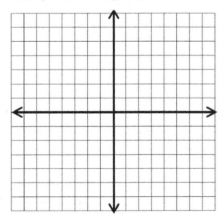

35) $f(x) = x, g(x) = x + 1$

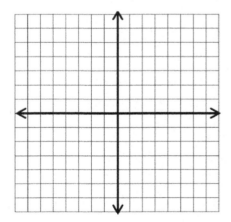

✍ **Sketch the group of each linear inequality.**

36) $y > 3x - 1$

37) $y < -x + 4$

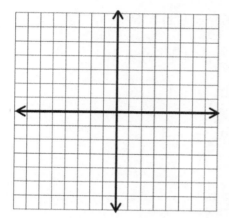

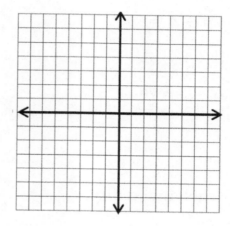

38) $y \leq -5x + 8$

39) $y \geq 2x - 1$

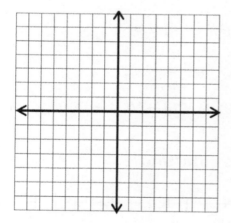

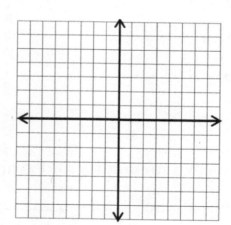

✍ **Solve each inequality.**

40) $2x - 8 \leq 6$

41) $4x - 21 < 19$

42) $8x - 2 \leq 14$

43) $2x - 3 < 21$

44) $-5 + 3x \leq 10$

45) $17 - 3x \geq -13$

46) $2(x - 3) \leq 6$

47) $2 - 3x > -7$

48) $7x - 5 \leq 9$

49) $3 + 2x \geq 19$

**Effortless**
**Math**
**Education**

✎ **Solve each equation.**

50) $|a| + 12 = 22$

51) $-2|x + 2| = -12$

52) $3|x + 4| = 45$

53) $|6x| + 4 = 70$

54) $-2|x| = -24$

55) $|-5m| = 30$

56) $|5 + x| = 5$

57) $|-4 + 5x| = 16$

58) $|-4x| = 16$

59) $\left|\dfrac{x}{4}\right| = 5$

✎ **Solve each inequality.**

60) $|x| + 4 \geq 6$

61) $|x - 2| - 6 < 5$

62) $3 + |2 + x| < 5$

63) $|x + 7| - 9 < -6$

64) $|x| - 3 > 2$

65) $|x| - 2 > 0$

66) $|3x| \leq 15$

67) $|x + 4| \leq 8$

68) $|3x| \leq 24$

69) $|x - 8| - 10 < -6$

✎ **Solve each inequality and graph its solution.**

70) $|2x - 2| \geq 10$

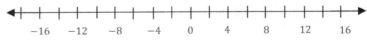

71) $\left|\dfrac{1}{3}x - 1\right| \leq 3$

72) $|x| - 2 < 6$

✎ **Solve each system of equations.**

73) $\begin{array}{l} -4x - 6y = 7 \\ x - 2y = 7 \end{array}$   $\begin{array}{l} x = \cdots \\ y = \cdots \end{array}$

76) $\begin{array}{l} -5x + y = -3 \\ 3x - 7y = 21 \end{array}$   $\begin{array}{l} x = \cdots \\ y = \cdots \end{array}$

74) $\begin{array}{l} 3y = -6x + 12 \\ 8x - 9y = -10 \end{array}$   $\begin{array}{l} x = \cdots \\ y = \cdots \end{array}$

77) $\begin{array}{l} x + 15y = 50 \\ x + 10y = 40 \end{array}$   $\begin{array}{l} x = \cdots \\ y = \cdots \end{array}$

75) $\begin{array}{l} 3x - 2y = 15 \\ 3x - 5y = 15 \end{array}$   $\begin{array}{l} x = \cdots \\ y = \cdots \end{array}$

78) $\begin{array}{l} 3x - 6y = -12 \\ -x - 3y = -6 \end{array}$   $\begin{array}{l} x = \cdots \\ y = \cdots \end{array}$

✎ **Determine if the following system of equations has a solution or not:**

79) $\begin{cases} 2y - 4x = 8 \\ y - 2x = 16 \end{cases}$

✎ **Solve each word problems.**

80) A theater is selling tickets for a performance. Mr. Smith purchased 8 senior tickets and 5 child tickets for $150 for his friends and family. Mr. Jackson purchased 4 senior tickets and 6 child tickets for $96. What is the price of a senior ticket? $_____

81) The difference of two numbers is 6. Their sum is 14. What is the bigger number? _____

82) The sum of the digits of a certain two-digit number is 7. Reversing its digits increase the number by 9. What is the number? _____

83) The difference of two numbers is 18. Their sum is 66. What are the numbers? _____

Effortless
Math
Education

# Chapter 2: Answers

1) $-3$

2) $-6$

3) $-12$

4) $-4$

5) $2$

6) $-1$

7) $16$

8) $-2$

9) $-5$

10) $-6$

11) $20$

12) $-1$

13) $2$

14) $1$

15) $-4$

16) $3$

17) $-1$

18) $3$

19) $2$

20) $-3$

21) $-1$

22) $3$

23) $2$

24) $-11$

25)

26)

27)

28)

29)

30) $x$ −intercepts: $(-4, 0)$, $y$ −intercepts: $(0, -1)$

31) $x$ −intercepts: $(8, 0)$, $y$ −intercepts: $(0, -12)$

32) $x$ −intercepts: $(5, 0)$, $y$ −intercepts: $(0, -3)$

33) $x$ −intercepts: $(\frac{5}{4}, 0)$, $y$ −intercepts: $(0, -5)$

34)

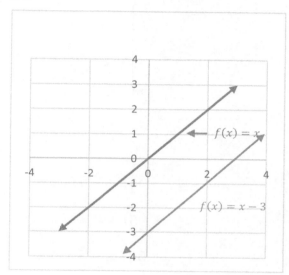

35)

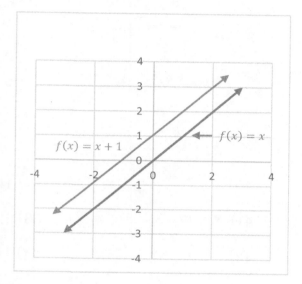

36) $y > 3x - 1$

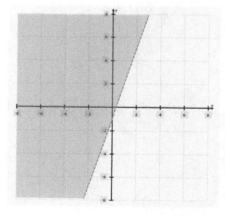

37) $y < -x + 4$

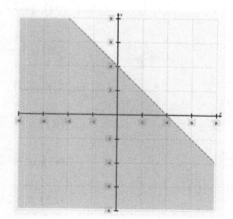

Effortless
Math
Education

38) $y \le -5x + 8$

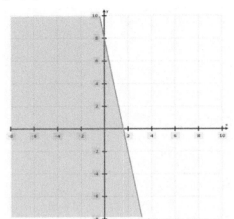

39) $y \ge 2x - 1$

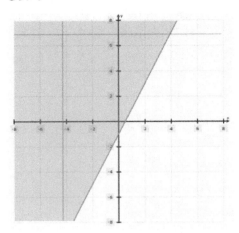

40) $x \le 7$

41) $x < 10$

42) $x \le 2$

43) $x < 12$

44) $x \le 5$

45) $x \le 10$

46) $x \le 6$

47) $x < 3$

48) $x \le 2$

49) $x \ge 8$

50) $\{10, -10\}$

51) $\{4, -8\}$

52) $\{11, -19\}$

53) $\{11, -11\}$

54) $\{-12, 12\}$

55) $\{6, -6\}$

56) $\{0, -10\}$

57) $\{-\frac{12}{5}, 4\}$

58) $\{4, -4\}$

59) $\{20, -20\}$

60) $x \ge 2$ or $x \le -2$

61) $x < 13$ and $x > -9$

62) $x < 0$ and $x > -4$

63) $x < -4$ and $x > -10$

64) $x > 5$ or $x < -5$

65) $x > 2$ or $x < -2$

66) $x \le 5$ and $x \ge -5$

67) $-12 \le x \le 4$

68) $x \le 8$ and $x \ge -8$

69) $4 < x < 12$

**Effortless Math Education**

70)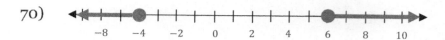

71)

72)

73) $x = 2, y = -\dfrac{5}{2}$

76) $x = 0, y = -3$

74) $x = 1, y = 2$

77) $x = 20, y = 2$

75) $x = 5, y = 0$

78) $x = 0, y = 2$

79) This system has no solution so
    it's a special system

80) $15

82) 34

81) 10

83) 42, 24

Effortless
Math
Education

# CHAPTER

## 3 Quadratic Function

Math topics that you'll learn in this chapter:

- ☑ Solving a Quadratic Equation
- ☑ Graphing Quadratic Functions
- ☑ Axis of Symmetry of Quadratic functions
- ☑ Solve a Quadratic Equation by Graphing
- ☑ Solving Quadratic Equation by Using Square Roots
- ☑ Build Quadratic from Roots
- ☑ Solving Quadratic Inequalities
- ☑ Graphing Quadratic Inequalities
- ☑ Factoring the Difference of Two Perfect Squares

# Solving a Quadratic Equation

- Write the equation in the form of: $ax^2 + bx + c = 0$.

- Factorize the quadratic, set each factor equal to zero and solve.

- Use quadratic formula if you couldn't factorize the quadratic.

- Quadratic formula: $x = \frac{-b \pm \sqrt{b^2 - 4ac}}{2a}$.

## Examples:

Find the solutions of each quadratic function.

**Example 1.** $x^2 + 7x + 12 = 0$.

*Solution:* Factor the quadratic by grouping. We need to find two numbers whose sum is 7 (from $7x$) and whose product is 12. Those numbers are 3 and 4. Then:

$x^2 + 7x + 12 = 0 \rightarrow x^2 + 3x + 4x + 12 = 0 \rightarrow (x^2 + 3x) + (4x + 12) = 0$.

Now, find common factors:

$(x^2 + 3x) = x(x + 3)$ and $(4x + 12) = 4(x + 3)$.

We have two expressions $(x^2 + 3x)$ and $(4x + 12)$ and their common factor is $(x + 3)$. Then:

$(x^2 + 3x) + (4x + 12) = 0 \rightarrow x(x + 3) + 4(x + 3) = 0 \rightarrow (x + 3)(x + 4) = 0$.

The product of two expressions is 0. Then:

$(x + 3) = 0 \rightarrow x = -3$ or $(x + 4) = 0 \rightarrow x = -4$.

**Example 2.** $x^2 + 5x + 6 = 0$.

*Solution:* Use quadratic formula:

$x_{1,2} = \frac{-b \pm \sqrt{b^2 - 4ac}}{2a}$, $a = 1$, $b = 5$ and $c = 6$.

We have $x = \frac{-5 \pm \sqrt{5^2 - 4 \times 1(6)}}{2(1)}$, therefore:

$$x_1 = \frac{-5 + \sqrt{5^2 - 4 \times 1(6)}}{2(1)} = -2,$$

$$x_2 = \frac{-5 - \sqrt{5^2 - 4 \times 1(6)}}{2(1)} = -3.$$

# Graphing Quadratic Functions

- Quadratic functions in vertex form: $y = a(x-h)^2 + k$ where $(h, k)$ is the vertex of the function. The axis of symmetry is $x = h$.

- Quadratic functions in standard form: $y = ax^2 + bx + c$ where $x = -\frac{b}{2a}$ is the value of $x$ in the vertex of the function.

- To graph a quadratic function, first find the vertex, then substitute some values for $x$ and solve for $y$. (Remember that the graph of a quadratic function is a $U$−shaped curve and it is called a "parabola".)

## Example:

Sketch the graph of $y = (x + 2)^2 - 3$.

**Solution**: Quadratic functions in vertex form:

$y = a(x-h)^2 + k$

Where $(h, k)$ is the vertex.

In addition, the axis of symmetry is:

$x = h$.

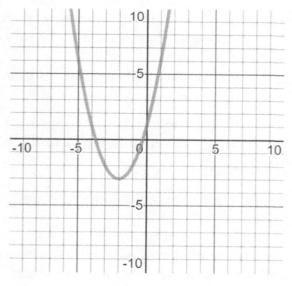

Then, the vertex of $y = (x + 2)^2 - 3$ is: $(-2, -3)$.

And the axis of symmetry is:

$x = -2$.

Now, substitute zero for $x$ and solve for $y$:

$y = (0 + 2)^2 - 3 = 1$.

The $y$−intercept is $(0,1)$.

Now, you can simply graph the quadratic function.

Notice that quadratic function is a $U$−shaped curve.

# Axis of Symmetry of Quadratic Functions

- An imaginary straight line that divides a figure into 2 equal parts is called the axis of symmetry.
- The quadratic function graph is a parabola. A parabola has one symmetry. The axis of symmetry is the straight line that splits a parabola into 2 identical parts.
- Parabolas are often in 4 forms. They are either horizontal, vertical, facing left, or facing right. The axis of symmetry specifies the type of parabola. The parabola is vertical and opens up or down if the axis of symmetry is vertical. The parabola is horizontal and opens left or right if the axis of symmetry is horizontal.
- The vertex of the parabola is the point that the axis of symmetry always passes through. The vertex's $x$ −coordinate is the equation of the parabola's axis of symmetry.
- When the quadratic equation is in standard form ($y = ax^2 + bx + c$), the axis of the symmetry formula is: $x = -\frac{b}{2a}$.
- When the quadratic equation is in vertex form ($y = a(x - h)^2 + k$), the axis of the symmetry formula is $x = h$.

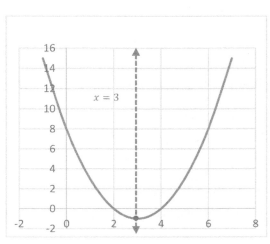

## Example:

Graph the following quadratic equation, then find the axis of symmetry of the quadratic function.

$$x^2 - 6x + 8$$

**Solution**: We know when the quadratic equation is in standard form ($y = ax^2 + bx + c$), the axis of symmetry formula is: $x = -\frac{b}{2a}$. In the equation $x^2 - 6x + 8$, $a = 1$, and $b = -6$. So, put $a = 1$, and $b = -6$ in the axis of symmetry formula:

$$x = -\frac{b}{2a} \rightarrow x = -\frac{-6}{2 \times 1} = \frac{6}{2} = 3.$$

The axis of symmetry is $x = 3$.

Find more at

# Solve a Quadratic Equation by Graphing

- A quadratic equation is a kind of equation that you can write in the form $ax^2 + bx + c = 0$. In this equation, $a$, $b$, and $c$ are numbers, and $a \neq 0$.

- To solve a quadratic equation by graphing, you can follow these steps:

  - 1st step: Make 2 new equations, each in two variables. To do this, set $y$ equal to the left side of the quadratic equation and then set $y$ equal to the right side of the quadratic equation.

  - 2nd step: Now you have obtained two new equations. Graph these 2 equations on a coordinate system and find the intersection points. All points that the graphs cross are named intersection points.

  - 3rd step: The $x$ −coordinates of all intersection points that you found are the solutions to the first quadratic equation. In fact, it's at these $x$ values that the left side of the quadratic equation equals the right side of the quadratic equation, and this is exactly what it represents to be an answer to the quadratic equation.

## Example:

Solve the following quadratic equation using graphing.
$$x^2 + 2x - 3 = 0$$

*Solution*: First, make 2 new equations: $y = x^2 + 2x - 3$ and $y = 0$. Then graph these 2 equations on a coordinate system and find the intersection points. Here intersection points are $x = -3$ and $x = 1$. At these two points the value of $y$ in the 2 equations is equal ($y = 0$). So, $x = -3$ and $x = 1$ are the solutions to the first quadratic equation.

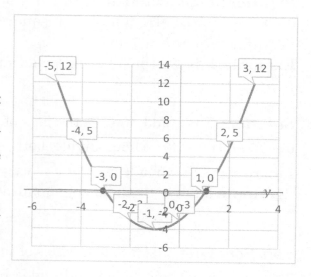

bit.ly/3GlTTAj

Find more at

# Solving Quadratic Equations by Using Square Roots

- Using the square root is one way that you can use to solve quadratic equations. This technique is usually used for equations that are in form $ax^2 = c$ or $(ax + b)^2 = c$, or an equation that is rewritten in either of these two forms.

- For solving an equation by using the square root; in the first step, the term that includes the squared variable should be isolated. In the next step, you should get the square root of both sides and then solve for the variable. In the last step, you should write the final answer as simply as possible.

- Keep in mind that there are always 2 roots for every square root: one of them is positive and another one is negative. After you take the square root, put a $\pm$ sign before the two possible roots and make sure that the final answer includes both possible roots.

## Examples:

**Example 1.** Solve the following quadratic equation using the Square Root.

$$x^2 - 81 = 0$$

**Solution**: First add 81 to both sides of the quadratic equation: $x^2 - 81 + 81 = 0 + 81 \rightarrow x^2 = 81$. Then, take the square root of both sides: $x = \pm\sqrt{81} \rightarrow x = \pm 9$. So, the solutions are 9 and $-9$.

**Example 2.** Solve the following quadratic equation using the Square Root.

$$9x^2 - 36 = 0$$

**Solution**: First, add 36 to both sides of the quadratic equation: $9x^2 - 36 + 36 = 0 + 36 \rightarrow 9x^2 = 36$. Then, divide both sides by 9: $\frac{9x^2}{9} = \frac{36}{9} \rightarrow x^2 = 4$. Take the square root of both sides: $x = \pm\sqrt{4} \rightarrow x = \pm 2$. So, the solutions are 2 and $-2$.

**Example 3.** Solve the following quadratic equation using the Square Root.

$$6x^2 - 3 = 51$$

**Solution:** First, add 3 to both sides of the quadratic equation: $6x^2 - 3 + 3 = 51 + 3 \rightarrow 6x^2 = 54$. Then divide both sides by 6: $\frac{6x^2}{6} = \frac{54}{6} \rightarrow x^2 = 9$. Finally, take the square root of both sides: $x = \pm\sqrt{9} \rightarrow x = \pm 3$. So, the solutions are 3 and $-3$.

# Build Quadratics from Roots

- To construct a quadratic equation from its root, if you have rational solutions, you can use the reverse factoring method. First, you can make each of the solutions equal to $x$. In the next step, make the equation equal to 0 using an addition or subtraction operation. When you've passed this step, you have succeeded in converting the roots to quadratic factors.

- The formula to build the quadratic equation from its root is $x^2 - (\alpha + \beta)x + \alpha\beta = 0$. Here, $\alpha$, and $\beta$ are the quadratic equation's roots. In other words, you can interpret this formula like this:

$$x^2 - (\text{sum of roots})x + (\text{product of roots}) = 0.$$

- If you have a quadratic equation in standard form $(ax^2 + bx + c = 0)$, you can use the coefficient of $x^2$, $x$, and constant term and find the root's sum and product. Here you can consider $\alpha$ and $\beta$ as the two roots of the quadratic equation. In this case, the formula to find the sum and product of the quadratic equation's root is as follows:

$$\alpha + \beta = -\frac{b}{a} = -\frac{coefficient\ of\ x}{coefficient\ of\ x^2}$$

$$\alpha\beta = \frac{c}{a} = \frac{constant\ term}{coefficient\ of\ x^2}$$

- Note that the roots of a quadratic equation always happen in conjugate pairs.

## Examples:

**Example 1.** Build the quadratic equation whose roots are 3 and 5.
**Solution**: 3 and 5 are $\alpha$, and $\beta$. So first find the value of $\alpha + \beta$: $3 + 5 = 8$. Now find the value of $\alpha\beta$: $3 \times 5 = 15$. Put both of these values in $x^2 - (\alpha + \beta)x + \alpha\beta = 0$: $x^2 - 8x + 15 = 0$.

**Example 2.** Build the quadratic equation whose roots are $\frac{3}{2}$ and $\frac{1}{4}$.
**Solution**: $\frac{3}{2}$ and $\frac{1}{4}$ are $\alpha$, and $\beta$. So, first find the value of $\alpha + \beta$: $\frac{3}{2} + \frac{1}{4}$. The least common multiplication of the denominators 4 and 2 is 4. $\frac{3}{2} + \frac{1}{4} = \frac{6+1}{4} = \frac{7}{4}$. In the next step, find the value of $\alpha\beta$: $\frac{3}{2} \times \frac{1}{4} = \frac{3}{8}$. Put both of these values in $x^2 - (\alpha + \beta)x + \alpha\beta = 0$: $x^2 - \frac{7}{4}x + \frac{3}{8} = 0$. Now, Multiply each side by 8: $8x^2 - 14x + 3 = 0$.

# Solving Quadratic Inequalities

- A quadratic inequality is one that can be written in the standard form of $ax^2 + bx + c > 0$. (Or substitute $<$, $\leq$, or $\geq$ for $>$)

- Solving a quadratic inequality is like solving equations. We need to find the solutions (the zeroes).

- To solve quadratic inequalities, first, find quadratic equations. Then choose a test value between zeroes. Finally, find interval(s), such as $> 0$ or $< 0$.

## Examples:

**Example 1.** Solve quadratic inequality. $x^2 + x - 6 > 0$

*Solution*: First solve $x^2 + x - 6 = 0$ by factoring. Then:
$x^2 + x - 6 = 0 \rightarrow (x - 2)(x + 3) = 0$.
The product of two expressions is 0. Then:
$(x - 2) = 0 \rightarrow x = 2$ or $(x + 3) = 0 \rightarrow x = -3$.
Now, choose a value between 2 and $-3$. Let's choose 0. Then:
$x = 0 \rightarrow x^2 + x - 6 > 0 \rightarrow (0)^2 + (0) - 6 > 0 \rightarrow -6 > 0$.
$-6$ is not greater than 0. Therefore, all values between 2 and $-3$ are NOT the solution of this quadratic inequality.
The solution is: $x > 2$ or $x < -3$. To represent the solution, we can use interval notation, in which solution sets are indicated with parentheses or brackets. The solutions $x > 2$ or $x < -3$ are represented as: $(-\infty, -3) \cup (2, \infty)$.

**Example 2.** Solve quadratic inequality. $x^2 - 2x - 8 \geq 0$

*Solution*: First solve: $x^2 - 2x - 8 = 0$, Factor:
$x^2 - 2x - 8 = 0 \rightarrow (x - 4)(x + 2) = 0$.
$-2$ and 4 are the solutions. Choose a point between $-2$ and 4. Let's choose 0.
Then: $x = 0 \rightarrow x^2 - 2x - 8 \geq 0 \rightarrow (0)^2 - 2(0) - 8 \geq 0 \rightarrow -8 \geq 0$. This is NOT true.
So, the solution is: $x \leq -2$ or $x \geq 4$. (Using interval notation) The solution is: $(-\infty, -2] \cup [4, \infty)$.

# Graphing Quadratic Inequalities

- A quadratic inequality is in the form:

  $y > ax^2 + bx + c$ (Or substitute $<$, $\leq$ or $\geq$ for $>$).

- To graph a quadratic inequality, start by graphing the quadratic parabola. Then fill in the region either inside or outside of it, depending on the inequality.

- Choose a testing point and check the solution section.

## Example:

Sketch the graph of $y > 2x^2$.

*Solution*: First, graph the quadratic:

$y = 2x^2$.

Quadratic functions in vertex form:

$y = a(x-h)^2 + k$

Where $(h, k)$ is the vertex.

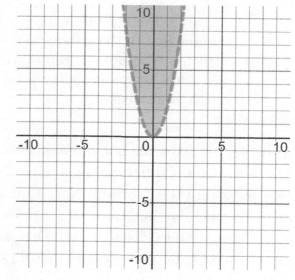

Then, the vertex of $y = 2x^2$ is:

$(h, k) = (0,0)$.

Since the inequality sing is $>$, we need to use dash lines.

Now, choose a testing point inside the parabola. Let's choose $(0, 2)$.

$y > 2x^2 \rightarrow 2 > 2(0)^2 \rightarrow 2 > 0$.

This is true. So, inside the parabola is the solution section.

# Factoring the Difference of Two Perfect Squares

- Factoring the difference of two perfect squares technique is used when you want to factor a polynomial that includes the two perfect squares subtraction. The formula for factoring the difference between two perfect squares is $a^2 - b^2 = (a - b)(a + b)$ and using this formula you can easily find the square root of every perfect square in the polynomial expression. Then you can put obtained values into the formula. In algebra, the difference between two perfect squares technique is a good method that you can use generally when you want to solve equations related to perfect-squares.

- To factor the difference between two perfect squares, you can take these steps:
  - 1st step: Use the difference of squares formula. $a^2 - b^2 = (a - b)(a + b)$. $a^2$ and $b^2$ are the terms of the perfect squares in your expression, and $a$ and $b$ are the perfect squares roots.
  - 2nd step: Put the value for $a$ into the formula. You should find this value, so, consider the first perfect square in the polynomial expression and find its square root. Note that the factor that you multiply by itself to find a number is that number's square root.
  - 3rd step: Put the value for $b$ into the formula. This is the second term's square root in the polynomial expression.
  - 4th step: You can use the FOIL technique (multiplying binomials using the distributive property method) to multiply 2 obtained factors and check your work.

## Examples:

**Example 1.** Factor the polynomial. $144x^4 - 100$

**Solution**: Use the difference of squares formula: $a^2 - b^2 = (a - b)(a + b)$. Put the value for $a$ into the formula. Note that the factor that you multiply by itself to find a number is that number's square root: $144x^4 \rightarrow (12x^2)(12x^2) \rightarrow 144x^4 - 100 = (12x^2 - b)(12x^2 + b)$. Put the value for $b$ into the formula: $100 \rightarrow (10)(10) \rightarrow 144x^4 - 100 = (12x^2 - 10)(12x^2 + 10)$. The answer is $144x^4 - 100 = (12x^2 - 10)(12x^2 + 10)$.

**Example 2.** Factor the polynomial. $81x^2 - 7$

**Solution**: Use the difference of squares formula: $a^2 - b^2 = (a - b)(a + b)$. Put the value for $a$ into the formula: $81x^2 \rightarrow (9x)(9x) \rightarrow 81x^2 - 7 = (9x - b)(9x + b)$. Put the value for $b$ into the formula:

$$7 \rightarrow (\sqrt{7})(\sqrt{7}) \rightarrow 81x^2 - 7 = (9x - \sqrt{7})(9x + \sqrt{7}).$$
The answer is $81x^2 - 7 = (9x - \sqrt{7})(9x + \sqrt{7})$.

# Chapter 3: Practices

✍ **Solve each equation.**

1) $x^2 - 5x - 14 = 0$

2) $x^2 + 8x + 15 = 0$

3) $x^2 - 5x - 36 = 0$

4) $x^2 - 12x + 35 = 0$

5) $x^2 + 12x + 32 = 0$

6) $5x^2 + 27x + 28 = 0$

7) $8x^2 + 26x + 15 = 0$

8) $3x^2 + 10x + 8 = 0$

9) $12x^2 + 30x + 12 = 0$

10) $9x^2 + 57x + 18 = 0$

✍ **Sketch the graph of each function. Identify the vertex and axis of symmetry.**

11) $y = 3(x - 5)^2 - 2$

12) $y = x^2 - 7x + 15$

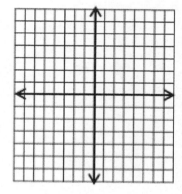
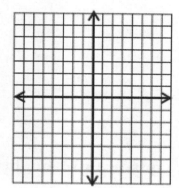

✍ **Graph the following quadratic equation, then find the axis of symmetry of the quadratic function.**

13) $x^2 - 4x + 6$

14) $x^2 + 8x + 12$

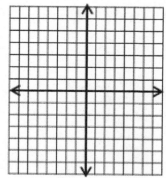
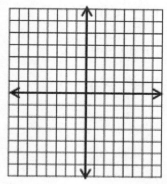

✎ **Solve the following quadratic equation using graphing.**

15) $x^2 + 4x - 5 = 0$          16) $x^2 - 8x + 15 = 0$

✎ **Solve.**

17) $x^2 - 16 = 0$          21) $2x^2 - 162 = 0$

18) $x^2 - 36 = 0$          22) $7x^2 - 448 = 0$

19) $x^2 - 32 = 0$          23) $9x^2 + 10 = 91$

20) $x^2 - 8 = 0$          24) $8x^2 - 8 = 192$

✎ **Build the quadratic equation whose roots are:**

25) $4, -2$          29) $-8, -3$

26) $3, 6$          30) $2, 9$

27) $-2, -5$          31) $-7, 6$

28) $6, -4$          32) $5, 9$

✎ **Solve each quadratic inequality.**

33) $x^2 + 7x + 10 < 0$          39) $x^2 - 16x + 64 \geq 0$

34) $x^2 + 9x + 20 > 0$          40) $x^2 - 36 \leq 0$

35) $x^2 - 8x + 16 > 0$          41) $x^2 - 13x + 36 \geq 0$

36) $x^2 - 8x + 12 \leq 0$          42) $x^2 + 15x + 36 \leq 0$

37) $x^2 - 11x + 30 \leq 0$          43) $4x^2 - 6x - 9 > x^2$

38) $x^2 - 12x + 27 \geq 0$          44) $5x^2 - 15x + 10 < 0$

**Effortless**
**Math**
**Education**

✎ **Sketch the graph of each function.**

45) $y > -2x^2$

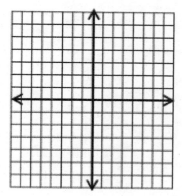

46) $y \geq 4x^2$

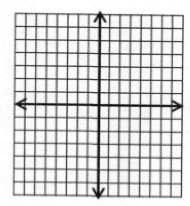

✎ **Factor the polynomials.**

47) $36x^4 - 25$

48) $81x^4 - 121$

49) $25x^4 - 64$

50) $49x^4 - 144$

51) $169x^4 - 121$

52) $81x^4 - 5$

**Effortless Math Education**

# Chapter 3: Answers

1) $x = -2, x = 7$

2) $x = -3, x = -5$

3) $x = 9, x = -4$

4) $x = 7, x = 5$

5) $x = -4, x = -8$

6) $x = -\frac{7}{5}, x = -4$

7) $x = -\frac{5}{2}, x = -\frac{3}{4}$

8) $x = -\frac{4}{3}, x = -2$

9) $x = -\frac{1}{2}, x = -2$

10) $x = -\frac{1}{3}, x = -6$

11)

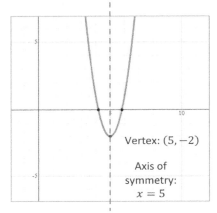

Vertex: $(5, -2)$

Axis of symmetry: $x = 5$

12)

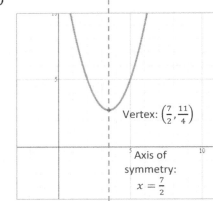

Vertex: $\left(\frac{7}{2}, \frac{11}{4}\right)$

Axis of symmetry: $x = \frac{7}{2}$

13) $x = 2$

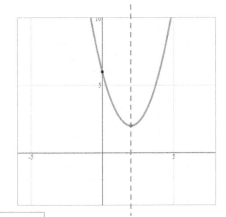

14) $x = -4$

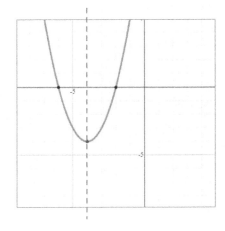

15) $x = 1, x = -5$

16) $x = 3, x = 5$

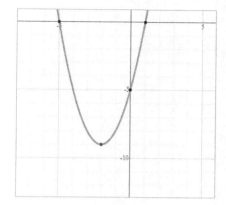

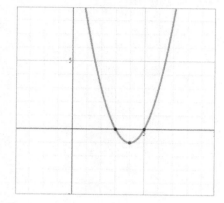

17) $x = \pm 4$

18) $x = \pm 6$

19) $x = \pm 4\sqrt{2}$

20) $x = \pm 2\sqrt{2}$

21) $x = \pm 9$

22) $x = \pm 8$

23) $x = \pm 3$

24) $x = \pm 5$

25) $x^2 - 2x - 8 = 0$

26) $x^2 - 9x + 18 = 0$

27) $x^2 + 7x + 10 = 0$

28) $x^2 - 2x - 24 = 0$

29) $x^2 + 11x + 24 = 0$

30) $x^2 - 11x + 18 = 0$

31) $x^2 + x - 42 = 0$

32) $x^2 - 14x + 45 = 0$

33) $-5 < x < -2$

34) $x < -5 \text{ or } x > -4$

35) $x < 4 \text{ or } x > 4$

36) $2 \leq x \leq 6$

37) $5 \leq x \leq 6$

38) $x \leq 3 \text{ or } x \geq 9$

**Effortless Math Education**

39) All real numbers

40) $-6 \leq x \leq 6$

41) $x \leq 4 \text{ or } x \geq 9$

42) $-12 \leq x \leq -3$

43) $x < -1 \text{ or } x > 3$

44) $1 < x < 2$

45)

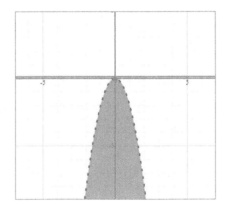

46)

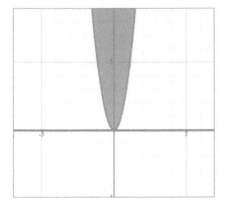

47) $(6x^2 - 5)(6x^2 + 5)$

48) $(9x^2 - 11)(9x^2 + 11)$

49) $(5x^2 - 8)(5x^2 + 8)$

50) $(7x^2 - 12)(7x^2 + 12)$

51) $(13x^2 - 11)(13x^2 + 11)$

52) $(9x^2 - \sqrt{5})(9x^2 + \sqrt{5})$

# 4 Complex Numbers

Math topics that you'll learn in this chapter:

- ☑ Adding and Subtracting Complex Numbers
- ☑ Multiplying and Dividing Complex Numbers
- ☑ Rationalizing Imaginary Denominators

51

# Adding and Subtracting Complex Numbers

- A complex number is expressed in the form $a + bi$, where $a$ and $b$ are real numbers, and $i$, which is called an imaginary number, is a solution of the equation $i^2 = -1$.

- For adding complex numbers:

$$(a + bi) + (c + di) = (a + c) + (b + d)i$$

- For subtracting complex numbers:

$$(a + bi) - (c + di) = (a - c) + (b - d)i$$

## Examples:

**Example 1.** Solve: $(8 + 4i) + (6 - 2i)$.

*Solution*: Remove parentheses:

$(8 + 4i) + (6 - 2i) = 8 + 4i + 6 - 2i$.

Combine like terms:

$8 + 4i + 6 - 2i = 14 + 2i$.

**Example 2.** Solve: $(10 + 8i) + (8 - 3i)$.

*Solution*: Remove parentheses:

$(10 + 8i) + (8 - 3i) = 10 + 8i + 8 - 3i$.

Group like terms:

$10 + 8i + 8 - 3i = 18 + 5i$.

**Example 3.** Solve: $(-5 - 3i) - (2 + 4i)$.

*Solution*: Remove parentheses by multiplying $-1$ to the second parentheses:

$(-5 - 3i) - (2 + 4i) = -5 - 3i - 2 - 4i$.

Combine like terms:

$-5 - 3i - 2 - 4i = -7 - 7i$.

# Multiplying and Dividing Complex Numbers

- You can use FOIL (First-Out-In-Last) method or the following rule to multiply imaginary numbers. Remember that: $i^2 = -1$.

$$(a + bi) \cdot (c + di) = (ac - bd) + (ad + bc)i$$

- To divide complex numbers, you need to find the conjugate of the denominator. the conjugate of $(a + bi)$ is $(a - bi)$.

- Dividing complex numbers: $\frac{a+bi}{c+di} = \frac{a+bi}{c+di} \times \frac{c-di}{c-di} = \frac{ac+bd}{c^2+d^2} + \frac{bc-ad}{c^2+d^2}i$.

## Examples:

**Example 1.** Solve: $\frac{6-2i}{2+i}$.

**Solution**: The conjugate of $(2 + i)$ is $(2 - i)$. Use the rule for dividing complex numbers:
$\frac{a+bi}{c+di} = \frac{a+bi}{c+di} \times \frac{c-di}{c-di} = \frac{ac+bd}{c^2+d^2} + \frac{bc-ad}{c^2+d^2}i$.

Therefore:

$\frac{6-2i}{2+i} \times \frac{2-i}{2-i} = \frac{6\times(2)+(-2)(1)}{2^2+(1)^2} + \frac{-2\times2-(6)(1)}{2^2+(1)^2}i = \frac{10}{5} + \frac{-10}{5}i = 2 - 2i$.

**Example 2.** Solve: $(2 - 3i)(6 - 3i)$.

**Solution**: Use the multiplication of imaginary numbers rule:
$(a + bi) \cdot (c + di) = (ac - bd) + (ad + bc)i$.
Therefore:
$(2 \times 6 - (-3)(-3)) + (2(-3) + (-3) \times 6)i = 3 - 24i$.

**Example 3.** Solve: $\frac{3-2i}{4+i}$.

**Solution**: Use the rule for dividing complex numbers:

$\frac{a+bi}{c+di} = \frac{a+bi}{c+di} \times \frac{c-di}{c-di} = \frac{ac+bd}{c^2+d^2} + \frac{bc-ad}{c^2+d^2}i$.

Therefore:

$\frac{3-2i}{4+i} \times \frac{4-i}{4-i} = \frac{(3\times4+(-2i)\times(-i))+(-2\times4-3\times1)i}{4^2-i^2} = \frac{10-11i}{17} = \frac{10}{17} - \frac{11}{17}i$.

# Rationalizing Imaginary Denominators

- Step 1: Find the conjugate. (It's the denominator with different sign between the two terms.)
- Step 2: Multiply the numerator and denominator by the conjugate.
- Step 3: Simplify if needed.

## Examples:

**Example 1.** Solve: $\frac{4-3i}{6i}$.

**Solution**: Multiply both numerator and denominator by $\frac{i}{i}$:

$\frac{4-3i}{6i} = \frac{4-3i}{6i} \times \frac{i}{i}$.

Therefore:

$\frac{4-3i}{6i} = \frac{(4-3i)(i)}{6i(i)} = \frac{(4)(i)-(3i)(i)}{6(i^2)} = \frac{4i-3i^2}{6(-1)} = \frac{4i-3(-1)}{-6} = \frac{4i}{-6} + \frac{3}{-6} = -\frac{1}{2} - \frac{2}{3}i$.

**Example 2.** Solve: $\frac{6i}{2-i}$.

**Solution**: Multiply both numerator and denominator by the conjugate $\frac{2+i}{2+i}$:

$\frac{6i}{2-i} = \frac{6i(2+i)}{(2-i)(2+i)}$.

Apply complex arithmetic rule: $(a+bi)(a-bi) = a^2 + b^2$.

Therefore: $2^2 + (-1)^2 = 5$, then:

$\frac{6i(2+i)}{(2-i)(2+i)} = \frac{-6+12i}{5} = -\frac{6}{5} + \frac{12}{5}i$.

**Example 3.** Solve: $\frac{8-2i}{2i}$.

**Solution**: Factor 2 from both sides: $\frac{8-2i}{2i} = \frac{2(4-i)}{2i}$, divide both sides by 2:

$$\frac{2(4-i)}{2i} = \frac{(4-i)}{i}$$

Multiply both numerator and denominator by $\frac{i}{i}$:

$$\frac{(4-i)}{i} = \frac{(4-i)}{i} \times \frac{i}{i} = \frac{(4i-i^2)}{i^2} = \frac{1+4i}{-1} = -1 - 4i.$$

# Chapter 4: Practices

✎ **Simplify.**

1) $(-4i) - (7 - 2i) =$

2) $(-3 - 2i) - (2i) =$

3) $(8 - 6i) + (-5i) =$

4) $(-3 + 6i) - (-9 - i) =$

5) $(-5 + 15i) - (-3 + 3i) =$

6) $(-14 + i) - (-12 - 11i) =$

7) $(-18 - 3i) + (11 + 5i) =$

8) $(-11 - 9i) - (-9 - 3i) =$

9) $-8 + (2i) + (-8 + 6i) =$

10) $(-2 - i)(4 + i) =$

11) $(2 - 2i)^2 =$

12) $(4 - 3i)(6 - 6i) =$

13) $(5 + 4i)^2 =$

14) $(4i)(-i)(2 - 5i) =$

15) $(2 - 8i)(3 - 5i) =$

16) $\dfrac{9i}{3-i} =$

17) $\dfrac{2+4i}{14+4i} =$

18) $\dfrac{5+6i}{-1+8i} =$

19) $\dfrac{-8-i}{-4-6i} =$

20) $\dfrac{-1+5i}{-8-7i} =$

21) $\dfrac{-2-9i}{-2+7i} =$

22) $\dfrac{-8}{-5i} =$

23) $\dfrac{-5}{-i} =$

24) $\dfrac{3}{5i} =$

25) $\dfrac{6}{-4i} =$

26) $\dfrac{-6-i}{-1+6i} =$

27) $\dfrac{-9-3i}{-3+3i} =$

28) $\dfrac{4i+1}{-1+3i} =$

29) $\dfrac{6-3i}{2-i} =$

30) $\dfrac{-5+2i}{2-3i} =$

# Chapter 4: Answers

1) $-7 - 2i$

2) $-3 - 4i$

3) $8 - 11i$

4) $6 + 7i$

5) $-2 + 12i$

6) $-2 + 12i$

7) $-7 + 2i$

8) $-2 - 6i$

9) $-16 + 8i$

10) $-7 - 6i$

11) $-8i$

12) $6 - 42i$

13) $9 + 40i$

14) $8 - 20i$

15) $-34 - 34i$

16) $-\frac{9}{10} + \frac{27}{10}i$

17) $\frac{11}{53} + \frac{12}{53}i$

18) $\frac{43}{65} - \frac{46}{65}i$

19) $\frac{19}{26} - \frac{11}{13}i$

20) $-\frac{27}{113} - \frac{47}{113}i$

21) $-\frac{59}{53} + \frac{32}{53}i$

22) $\frac{-8}{5}i$

23) $-5i$

24) $-\frac{3}{5}i$

25) $\frac{3}{2}i$

26) $i$

27) $1 + 2i$

28) $\frac{11}{10} - \frac{7}{10}i$

29) $3$

30) $-\frac{16}{13} - \frac{11}{13}i$

**Effortless Math Education**

# 5 Matrices

Math topics that you'll learn in this chapter:

- ☑ Using Matrices to Represent Data
- ☑ Adding and Subtracting Matrices
- ☑ Matrix Multiplication
- ☑ Solving Systems with Matrix Equations
- ☑ Finding Determinants of a Matrix
- ☑ The Inverse of a Matrix

57

# Using Matrices to Represent Data

- Matrices are two-dimensional arrays and are used to analyze numerical data of tables, linear equations and …

- A matrix is a rectangular array of numbers, which is represented as follows:

$$A = \begin{bmatrix} a_{11} & a_{12} & \cdots & a_{1n} \\ a_{21} & a_{22} & & a_{2n} \\ \vdots & & \ddots & \vdots \\ a_{m1} & a_{m2} & \cdots & a_{mn} \end{bmatrix}$$

Each number in the matrix is called an entry.

- The number of horizontal rows and the number of vertical columns of a matrix is called the dimension of it. The dimension of the matrix $A$ is $m \times n$.

- A matrix whose number of rows is equal to the number of columns is called a square matrix.

## Examples

**Example 1.** Solve $\begin{bmatrix} -1 & a+2 \\ 4 & 2 \\ 1 & -1 \end{bmatrix} = \begin{bmatrix} -1 & 0 \\ 2b-a & 2 \\ 1 & -1 \end{bmatrix}$ for $a$ and $b$.

*Solution*: Since two matrices are equal, each of the corresponding arrays are equal. Therefore: $a + 2 = 0 \rightarrow a = -2$.

In addition: $2b - a = 4$. So $a = -2$, then:

$2b - a = 4 \rightarrow 2b - (-2) = 4 \rightarrow 2b = 2 \rightarrow b = 1$.

**Example 2.** The following table shows the age and shoe size of each student in a class. Write the matrix corresponding to the table.
*Solution*: Put each student's number as a row of the matrix and age and shoe size as its column. Therefore, the matrix corresponding to the table is a matrix of dimension 5 × 2. As follow:

| Student | Size | Age |
|---------|------|-----|
| 1 | 37 | 11 |
| 2 | 38 | 12 |
| 3 | 27 | 10 |
| 4 | 39 | 10 |
| 5 | 35 | 11 |

$$S = \begin{bmatrix} 37 & 11 \\ 38 & 12 \\ 27 & 10 \\ 39 & 10 \\ 35 & 11 \end{bmatrix}$$

# Adding and Subtracting Matrices

- A matrix (plural: matrices) is a rectangular array of numbers or variables arranged in rows and columns.

- We can add or subtract two matrices if they have the same dimensions.

  For addition or subtraction, add or subtract the corresponding entries, and place the result in the corresponding position in the resultant matrix.

## Examples:

**Example 1.** $[1 \quad -4 \quad 6] + [2 \quad -3 \quad -9] =$

*Solution:* Add the elements in the matching positions:

$[1 + 2 \quad -4 + (-3) \quad 6 + (-9)] = [3 \quad -7 \quad -3].$

**Example 2.** $\begin{bmatrix} 2 & 4 \\ -5 & -1 \\ -2 & -6 \end{bmatrix} + \begin{bmatrix} 1 & 0 \\ 0 & 7 \\ 3 & 5 \end{bmatrix} =$

*Solution:* Add the elements in the matching positions:

$\begin{bmatrix} 2 + 1 & 4 + 0 \\ (-5) + 0 & (-1) + 7 \\ (-2) + 3 & (-6) + 5 \end{bmatrix} = \begin{bmatrix} 3 & 4 \\ -5 & 6 \\ 1 & -1 \end{bmatrix}.$

**Example 3.** $\begin{bmatrix} 1 & -1 \\ 2 & 0 \end{bmatrix} - \begin{bmatrix} 4 & 0 \\ 2 & -1 \end{bmatrix} =$

*Solution:* Subtract the elements in the matching positions:

$\begin{bmatrix} 1 - 4 & -1 - 0 \\ 2 - 2 & 0 - (-1) \end{bmatrix} = \begin{bmatrix} -3 & -1 \\ 0 & 1 \end{bmatrix}.$

**Example 4.** $\begin{bmatrix} -1 & 2 & 0 \\ 3 & -2 & 7 \\ 0 & 1 & -1 \end{bmatrix} - \begin{bmatrix} 0 & 2 & 4 \\ 1 & 0 & -1 \\ 1 & 0 & -1 \end{bmatrix} =$

*Solution:* Subtract the elements in the matching positions:

$\begin{bmatrix} -1 - 0 & 2 - 2 & 0 - 4 \\ 3 - 1 & -2 - 0 & 7 - (-1) \\ 0 - 1 & 1 - 0 & -1 - (-1) \end{bmatrix} = \begin{bmatrix} -1 & 0 & -4 \\ 2 & -2 & 8 \\ -1 & 1 & 0 \end{bmatrix}.$

# Matrix Multiplication

- Step 1: Make sure that it's possible to multiply the two matrices (The number of columns in the 1st one should be the same as the number of rows in the second one.)

- Step 2: The elements of each row of the first matrix should be multiplied by the elements of each column in the second matrix.

- Step 3: Add the products.

## Examples:

**Example 1.** $\begin{bmatrix} -1 & -3 \\ -4 & 0 \end{bmatrix} \begin{bmatrix} -3 & -2 \\ 4 & 4 \end{bmatrix} =$

*Solution:* Multiply the rows of the first matrix by the columns of the second matrix.

$$\begin{bmatrix} (-1)(-3)+(-3)(4) & (-1)(-2)+(-3)(4) \\ (-4)(-3)+(0)(4) & (-4)(-2)+(0)(4) \end{bmatrix} = \begin{bmatrix} -9 & -10 \\ 12 & 8 \end{bmatrix}.$$

**Example 2.** $\begin{bmatrix} -1 & -5 & -2 \\ 5 & 0 & 4 \end{bmatrix} \begin{bmatrix} 4 \\ 0 \\ 2 \end{bmatrix} =$

*Solution:* Multiply the rows of the first matrix by the columns of the second matrix.

$$\begin{bmatrix} (-1)(4)+(-5)(0)+(-2)(2) \\ (5)(4)+(0)(0)+(4)(2) \end{bmatrix} = \begin{bmatrix} -8 \\ 28 \end{bmatrix}.$$

**Example 3.** $\begin{bmatrix} 1 & 0 & -1 \\ 2 & -1 & 1 \end{bmatrix} \begin{bmatrix} 1 & 2 \\ 2 & -1 \\ 3 & 1 \end{bmatrix} =$

*Solution:* Multiply the rows of the first matrix by the columns of the second matrix.

$$\begin{bmatrix} (1)(1)+(0)(2)+(-1)(3) & (1)(2)+(0)(-1)+(-1)(1) \\ (2)(1)+(-1)(2)+(1)(3) & (2)(2)+(-1)(-1)+(1)(1) \end{bmatrix} = \begin{bmatrix} -2 & 1 \\ 3 & 6 \end{bmatrix}.$$

# Solving Systems with Matrix Equations

- Any system of linear equations can be written as a matrix equation, $AX = B$ where $A$ is the coefficient matrix, $X$ is the variable matrix and $B$ is the constant matrix.

- To solve a system of linear equations with a matrix equation, use the following steps:

  Step1: Rewrite the system of equations in standard form.

  Step2: Write the system as a matrix equation.

  Step3: Obtain the determinant of the matrix of coefficients and solve for $X = A^{-1}B$.

## Example

What is the value of $y$ in the following system of equations?
$$2x - 1 = 3y$$
$$y = 4 + x$$

*Solution*: First the system of equations in standard form. As follow:

$2x - 3y = 1$
$x - y = -4$ '

Write the system as a matrix equation:

$\begin{matrix} 2x - 3y = 1 \\ x - y = -4 \end{matrix} \rightarrow \begin{bmatrix} 2 & -3 \\ 1 & -1 \end{bmatrix} \begin{bmatrix} x \\ y \end{bmatrix} = \begin{bmatrix} 1 \\ -4 \end{bmatrix}.$

Let $A = \begin{bmatrix} 2 & -3 \\ 1 & -1 \end{bmatrix}$, obtain the determinant of $A$. Then, $|A| = 2 \times (-1) - (-3) \times 1 = 1$. Therefore:

$A^{-1} = \frac{1}{|A|} \begin{bmatrix} -1 & 3 \\ -1 & 2 \end{bmatrix} = \begin{bmatrix} -1 & 3 \\ -1 & 2 \end{bmatrix}.$

Solve $X = A^{-1}B$. It means that:

$X = \begin{bmatrix} x \\ y \end{bmatrix} = \begin{bmatrix} -1 & 3 \\ -1 & 2 \end{bmatrix} \begin{bmatrix} 1 \\ -4 \end{bmatrix} = \begin{bmatrix} (-1) \times 1 + 3 \times (-4) \\ (-1) \times 1 + 2 \times (-4) \end{bmatrix} = \begin{bmatrix} -13 \\ -9 \end{bmatrix}.$

Finally, the value of $y$ is equal to $-9$.

bit.ly/3GKQHzH

Find more at

# Finding Determinants of a Matrix

$$\begin{bmatrix} a & b \\ c & d \end{bmatrix} \qquad |A| = ad - bc$$

$$\begin{bmatrix} a & b & c \\ d & e & f \\ g & h & i \end{bmatrix} \qquad |A| = a(ei - fh) - b(di - fg) + c(dh - eg)$$

## Examples:

**Example 1.** Evaluate the determinant of matrix. $\begin{bmatrix} 1 & -2 \\ -5 & 0 \end{bmatrix}$

*Solution:* Use the matrix determinant:

$|A| = ad - bc.$

Therefore:

$|A| = (1)(0) - (-2)(-5) = -10.$

**Example 2.** Evaluate the determinant of matrix. $\begin{bmatrix} 2 & 6 & 3 \\ 0 & 5 & 1 \\ 4 & 7 & 4 \end{bmatrix}$

*Solution:* Use the matrix determinant:

$|A| = a(ei - fh) - b(di - fg) + c(dh - eg).$

Then:

$|A| = 2(5 \times 4 - 7 \times 1) - 6(0 \times 4 - 4 \times 1) + 3(0 \times 7 - 5 \times 4) = -10.$

**Example 3.** Evaluate the determinant of matrix. $\begin{bmatrix} 1 & 2 \\ -1 & -1 \end{bmatrix}$

*Solution:* Use the matrix determinant: $|A| = ad - bc.$

Therefore:

$|A| = (1)(-1) - (2)(-1) = 1.$

bit.ly/3jUX0HR

Find more at

# The Inverse of a Matrix

- For an arbitrary matrix $A$ with the following two conditions:
    - Be a square matrix.
    - Be a non-singular matrix. that's mean $|A| \neq 0$.

- There exists a matrix $B$ such that:
    - With matrix $A$ having the same dimension.
    - $BA = AB = I$, where $I$ is the identity matrix.

- The inverse of matrix $A$ is denoted by $A^{-1}$ $(A^{-1} \neq \frac{1}{A})$. For a $2 \times 2$ matrix:

$$A = \begin{bmatrix} a & b \\ c & d \end{bmatrix} \rightarrow A^{-1} = \frac{1}{|A|}\begin{bmatrix} d & -b \\ -c & a \end{bmatrix}$$

$$A = \begin{bmatrix} a_{11} & a_{12} & a_{13} \\ a_{21} & a_{22} & a_{23} \\ a_{31} & a_{32} & a_{33} \end{bmatrix} \rightarrow A^{-1} = \frac{1}{|A|}\begin{bmatrix} \begin{vmatrix} a_{22} & a_{23} \\ a_{32} & a_{33} \end{vmatrix} & \begin{vmatrix} a_{13} & a_{12} \\ a_{33} & a_{32} \end{vmatrix} & \begin{vmatrix} a_{12} & a_{13} \\ a_{22} & a_{23} \end{vmatrix} \\ \begin{vmatrix} a_{23} & a_{21} \\ a_{33} & a_{31} \end{vmatrix} & \begin{vmatrix} a_{11} & a_{13} \\ a_{31} & a_{33} \end{vmatrix} & \begin{vmatrix} a_{13} & a_{11} \\ a_{23} & a_{21} \end{vmatrix} \\ \begin{vmatrix} a_{21} & a_{22} \\ a_{31} & a_{32} \end{vmatrix} & \begin{vmatrix} a_{12} & a_{11} \\ a_{32} & a_{31} \end{vmatrix} & \begin{vmatrix} a_{11} & a_{12} \\ a_{21} & a_{22} \end{vmatrix} \end{bmatrix}$$

## Examples

**Example 1.** Show that $A = \begin{bmatrix} 2 & 1 \\ -1 & 0 \end{bmatrix}$ and $B = \begin{bmatrix} 0 & -1 \\ 1 & 2 \end{bmatrix}$ are inverses of one another.

*Solution*: For two matrices to be invertible, they must $BA = AB = I$. Therefore:

$$A \times B = \begin{bmatrix} 2 & 1 \\ -1 & 0 \end{bmatrix}\begin{bmatrix} 0 & -1 \\ 1 & 2 \end{bmatrix} = \begin{bmatrix} 2 \times 0 + 1 \times 1 & 2 \times (-1) + 1 \times 2 \\ (-1) \times 0 + 0 \times 1 & (-1) \times (-1) + 0 \times 2 \end{bmatrix} = \begin{bmatrix} 1 & 0 \\ 0 & 1 \end{bmatrix} = I,$$

$$B \times A = \begin{bmatrix} 0 & -1 \\ 1 & 2 \end{bmatrix}\begin{bmatrix} 2 & 1 \\ -1 & 0 \end{bmatrix} = \begin{bmatrix} 0 \times 2 + (-1) \times (-1) & 0 \times 1 + (-1) \times 0 \\ 1 \times 2 + 2 \times (-1) & 1 \times 1 + 2 \times 0 \end{bmatrix} = \begin{bmatrix} 1 & 0 \\ 0 & 1 \end{bmatrix} = I.$$

$A$ and $B$ are inverses of each other.

**Example 2.** Find the inverse of the matrix: $C = \begin{bmatrix} -1 & 3 \\ 0 & 2 \end{bmatrix}$.

*Solution*: Since $C$ is a square matrix, calculate the determinant of the matrix. Then: $|A| = \left|\begin{bmatrix} a & b \\ c & d \end{bmatrix}\right| = ad - bc \rightarrow |C| = (-1) \times 2 - 3 \times 0 = -2.$

Now, using this formula $A = \begin{bmatrix} a & b \\ c & d \end{bmatrix} \rightarrow A^{-1} = \frac{1}{|A|}\begin{bmatrix} d & -b \\ -c & a \end{bmatrix}$. We have:

$$C^{-1} = \frac{1}{|C|}\begin{bmatrix} 2 & -3 \\ 0 & -1 \end{bmatrix} \rightarrow C^{-1} = -\frac{1}{2}\begin{bmatrix} 2 & -3 \\ 0 & -1 \end{bmatrix} = \begin{bmatrix} -1 & \frac{3}{2} \\ 0 & \frac{1}{2} \end{bmatrix}.$$

# Chapter 5: Practices

✍ **Solve for *a* and *b*.**

1) $\begin{bmatrix} -1 & -2 \\ 2 & 6 \\ 1 & a+5 \end{bmatrix} = \begin{bmatrix} -1 & -2 \\ 2 & 2b-a \\ 1 & -1 \end{bmatrix}$

1) $\begin{bmatrix} 4 & -2 \\ 2 & a+4 \end{bmatrix} = \begin{bmatrix} 4 & 2a+b \\ 2 & a+4 \end{bmatrix} = \begin{bmatrix} 4 & 2a+b \\ 2 & 7 \end{bmatrix}$

✍ **Solve.**

3) $\begin{bmatrix} 2 & 1 \\ -1 & 3 \end{bmatrix} - \begin{bmatrix} 2 & 5 \\ -7 & -2 \end{bmatrix} =$

4) $\begin{bmatrix} 6 & 4 \\ -9 & 7 \end{bmatrix} + \begin{bmatrix} 5 & 3 \\ -4 & 1 \end{bmatrix} =$

5) $\begin{bmatrix} 2 & 0 \\ -1 & 1 \end{bmatrix} - \begin{bmatrix} 4 & -2 \\ 2 & 1 \end{bmatrix} =$

6) $\begin{bmatrix} 6 & -7 \\ -3 & 11 \end{bmatrix} + \begin{bmatrix} 10 & -11 \\ 12 & 18 \end{bmatrix} =$

7) $\begin{bmatrix} -1 & 2 & -1 \\ 2 & -1 & 0 \end{bmatrix} - \begin{bmatrix} 2 & -5 & -4 \\ 1 & 1 & -3 \end{bmatrix} =$

8) $\begin{bmatrix} 8 & 12 \\ 14 & 21 \end{bmatrix} + \begin{bmatrix} 8 & -15 \\ 10 & -7 \end{bmatrix} =$

9) $\begin{bmatrix} 12 \\ 9 \\ 5 \end{bmatrix} + \begin{bmatrix} 18 \\ -14 \\ 19 \end{bmatrix} =$

10) $\begin{bmatrix} 14 \\ -16 \\ 13 \\ 21 \end{bmatrix} + \begin{bmatrix} -16 \\ 8 \\ -5 \\ -18 \end{bmatrix} =$

✍ **Solve.**

11) $\begin{bmatrix} -1 & 0 & 3 \end{bmatrix} \begin{bmatrix} 1 \\ 2 \\ -1 \end{bmatrix} =$

12) $\begin{bmatrix} -1 \\ 6 \\ -6 \end{bmatrix} \begin{bmatrix} 8 & 5 & 4 \end{bmatrix} =$

13) $\begin{bmatrix} 0 & 2 \\ 2 & -1 \end{bmatrix} \begin{bmatrix} -2 & 1 \\ 1 & 4 \end{bmatrix} =$

14) $\begin{bmatrix} 2 & 4 & 3 \\ 4 & 3 & 2 \end{bmatrix} \begin{bmatrix} 4 & 3 \\ 5 & 5 \\ 2 & 5 \end{bmatrix} =$

15) $\begin{bmatrix} 2 & 5 \\ -4 & -3 \end{bmatrix} \begin{bmatrix} 1 & -5 \\ 3 & 2 \end{bmatrix} =$

16) $\begin{bmatrix} 1 & -2 \\ -4 & 5 \end{bmatrix} \begin{bmatrix} 4 & 3 \\ 4 & 0 \end{bmatrix} =$

17) $\begin{bmatrix} 3 & 1 & 2 \\ -5 & 6 & 5 \end{bmatrix} \begin{bmatrix} 3 \\ 5 \\ 2 \end{bmatrix} =$

18) $\begin{bmatrix} -1 & 2 & 5 \\ 0 & -2 & -1 \end{bmatrix} \begin{bmatrix} 5 & 1 \\ 2 & -2 \\ 0 & 1 \end{bmatrix} =$

✍ **Find the inverse of the matrix.**

19) $C = \begin{bmatrix} -1 & -4 \\ 0 & -2 \end{bmatrix}$

21) $C = \begin{bmatrix} -2 & 1 \\ 4 & 2 \end{bmatrix}$

20) $C = \begin{bmatrix} 2 & 3 & -4 \\ 0 & -1 & -2 \\ 2 & 4 & 0 \end{bmatrix}$

22) $C = \begin{bmatrix} -2 & 1 & -1 \\ 0 & 4 & -4 \\ 0 & 2 & -4 \end{bmatrix}$

✍ **What is the value of $x$ and $y$ in the following system of equations?**

23) $2x + 2y = 14$

$-10x - 2y = -54$

24) $-2x + 8y = -6$

$-2x + 4y = -6$

✍ **Evaluate the determinant of each matrix.**

25) $\begin{bmatrix} 2 & 3 \\ -7 & -1 \end{bmatrix} =$

30) $\begin{vmatrix} 2 & 9 & -1 \\ -1 & 4 & -2 \\ 1 & -4 & 1 \end{vmatrix} =$

26) $\begin{bmatrix} 3 & 1 & 5 \\ -1 & -4 & 1 \\ 5 & 3 & 0 \end{bmatrix} =$

31) $\begin{bmatrix} -6 & 12 \\ 3 & 0 \end{bmatrix} =$

27) $\begin{bmatrix} 4 & 6 \\ 8 & 1 \end{bmatrix} =$

32) $\begin{vmatrix} 3 & -4 & 1 \\ 4 & 2 & -8 \\ 6 & -3 & -2 \end{vmatrix} =$

28) $\begin{vmatrix} 9 & 2 & -1 \\ 3 & -1 & -5 \\ 2 & -2 & 1 \end{vmatrix} =$

33) $\begin{bmatrix} -1 & 1 & 7 \\ 4 & -2 & 7 \\ 0 & 2 & 1 \end{bmatrix} =$

29) $\begin{bmatrix} -3 & 9 \\ 4 & 5 \end{bmatrix} =$

34) $\begin{vmatrix} 6 & 5 & 1 \\ 1 & -4 & 2 \\ 7 & -2 & -2 \end{vmatrix} =$

Effortless
Math
Education

# Chapter 5: Answers

1) $a = -6$ and $b = 0$

2) $a = 3$ and $b = -8$

3) $\begin{bmatrix} 0 & -4 \\ 6 & 5 \end{bmatrix}$

4) $\begin{bmatrix} 11 & 7 \\ -13 & 8 \end{bmatrix}$

5) $\begin{bmatrix} -2 & 2 \\ -3 & 0 \end{bmatrix}$

6) $\begin{bmatrix} 16 & -18 \\ 9 & 29 \end{bmatrix}$

7) $\begin{bmatrix} -3 & 7 & 3 \\ 1 & -2 & 3 \end{bmatrix}$

8) $\begin{bmatrix} 16 & -3 \\ 24 & 14 \end{bmatrix}$

9) $\begin{bmatrix} 30 \\ -5 \\ 24 \end{bmatrix}$

10) $\begin{bmatrix} -2 \\ -8 \\ 8 \\ 3 \end{bmatrix}$

11) $[-4]$

12) $\begin{bmatrix} -8 & -5 & -4 \\ 48 & 30 & 24 \\ -48 & -30 & -24 \end{bmatrix}$

13) $\begin{bmatrix} 2 & 8 \\ -5 & -2 \end{bmatrix}$

14) $\begin{bmatrix} 34 & 41 \\ 35 & 37 \end{bmatrix}$

15) $\begin{bmatrix} 17 & 0 \\ -13 & 14 \end{bmatrix}$

16) $\begin{bmatrix} -4 & 3 \\ 4 & -12 \end{bmatrix}$

17) $\begin{bmatrix} 18 \\ 25 \end{bmatrix}$

18) $\begin{bmatrix} -1 & 0 \\ -4 & 3 \end{bmatrix}$

19) $C^{-1} = \begin{bmatrix} -1 & 2 \\ 0 & -\frac{1}{2} \end{bmatrix}$

20) $C^{-1} = \begin{bmatrix} -2 & 4 & \frac{5}{2} \\ 1 & -2 & -1 \\ -\frac{1}{2} & \frac{1}{2} & \frac{1}{2} \end{bmatrix}$

21) $C^{-1} = \begin{bmatrix} -\frac{1}{4} & \frac{1}{8} \\ \frac{1}{2} & \frac{1}{4} \end{bmatrix}$

22) $C^{-1} = \begin{bmatrix} -\frac{1}{2} & \frac{1}{8} & 0 \\ 0 & \frac{1}{2} & -\frac{1}{2} \\ 0 & \frac{1}{4} & -\frac{1}{2} \end{bmatrix}$

23) $x = 5$ and $y = 2$

24) $x = 3$ and $y = 0$

25) 19

26) 81

27) −44

28) −121

29) −51

30) −17

31) −36

32) 52

33) 68

34) 178

# CHAPTER

# 6 Polynomial Operations

Math topics that you'll learn in this chapter:

- ☑ Writing Polynomials in Standard Form
- ☑ Simplifying Polynomials
- ☑ Adding and Subtracting Polynomials
- ☑ Multiplying and Dividing Monomials
- ☑ Multiplying a Polynomial and a Monomial
- ☑ Multiplying Binomials
- ☑ Factoring Trinomials
- ☑ Choosing a Factoring Method for Polynomials
- ☑ Factoring by GCF
- ☑ Factors and Greatest Common Factors
- ☑ Operations with Polynomials
- ☑ Even and Odd Functions
- ☑ End Behavior of Polynomial Functions
- ☑ Remainder and Factor Theorems
- ☑ Polynomial Division (Long Division)
- ☑ Polynomial Division (Synthetic Division)
- ☑ Finding Zeros of Polynomials
- ☑ Polynomial Identities

# Writing Polynomials in Standard Form

- A polynomial function $f(x)$ of degree $n$ is of the form:

$$f(x) = a_n x^n + a_{n-1} x^{n-1} + \cdots + a_1 x + a_0$$

- The first term is the one with the biggest power!

## Examples:

**Example 1.** Write this polynomial in standard form. $8 + 5x^2 - 3x^3 =$

*Solution:* The highest exponent is the 3, so the entire term must be written first: $-3x^3$.

The next highest exponent is the 2, so that term comes next. So, we have: $-3x^3 + 5x^2$.

The constant term always comes last so the final answer is: $-3x^3 + 5x^2 + 8$.

**Example 2.** Write this polynomial in standard form. $5x^2 - 9x^5 + 8x^3 - 11 =$

*Solution:* The first term is the one with the biggest power: $5x^2 - 9x^5 + 8x^3 - 11 = -9x^5 + 8x^3 + 5x^2 - 11$.

**Example 3.** Write this polynomial in standard form.
$$7x^2 - 2x^6 - x^3 - 11 + 3x =$$

*Solution:* The first term is the one with the biggest power: $7x^2 - 2x^6 - x^3 - 11 + 3x = -2x^6 - x^3 + 7x^2 + 3x - 11$.

**Example 4.** Write this polynomial in standard form. $1 - 4a =$

*Solution:* The first term is the one with the biggest power: $1 - 4a = -4a + 1$.

**Example 5.** Write this polynomial in standard form. $-2 + 5y^3 - 4y =$

*Solution:* The first term is the one with the biggest power: $-2 + 5y^3 - 4y = 5y^3 - 4y - 2$.

# Simplifying Polynomials

- To simplify Polynomials, find "like" terms. (They have the same variables with the same power).

- Use "FOIL". (First–Out–In–Last) for binomials:

$$(x + a)(x + b) = x^2 + (b + a)x + ab$$

- Add or subtract "like" terms using order of operation.

## Examples:

**Example 1.** Simplify this expression. $x(4x + 7) - 2x =$

*Solution*: Use Distributive Property:
$x(4x + 7) = 4x^2 + 7x$.
Now, combine like terms:
$x(4x + 7) - 2x = 4x^2 + 7x - 2x = 4x^2 + 5x$.

**Example 2.** Simplify this expression. $(x + 3)(x + 5) =$

*Solution*: First, apply the FOIL method:
$(a + b)(c + d) = ac + ad + bc + bd$.
Now, we have:
$(x + 3)(x + 5) = x^2 + 5x + 3x + 15$.
Now combine like terms:
$x^2 + 5x + 3x + 15 = x^2 + 8x + 15$.

**Example 3.** Simplify this expression. $2x(x - 5) - 3x^2 + 6x =$

*Solution*: Use Distributive Property:
$2x(x - 5) = 2x^2 - 10x$.
Then:
$2x(x - 5) - 3x^2 + 6x = 2x^2 - 10x - 3x^2 + 6x$.
Now combine like terms:
$2x^2 - 3x^2 = -x^2$, and $-10x + 6x = -4x$.
The simplified form of the expression:
$2x^2 - 10x - 3x^2 + 6x = -x^2 - 4x$.

bit.ly/3rnAcj8

Find more at

# Adding and Subtracting Polynomials

- Adding polynomials is just a matter of combining like terms, with some order of operations considerations thrown in.

- Be careful with the minus signs, and don't confuse addition and multiplication!

- For subtracting polynomials, sometimes you need to use the Distributive Property: $a(b + c) = ab + ac$, $a(b - c) = ab - ac$.

## Examples:

**Example 1.** Simplify the expressions. $(x^2 - 2x^3) - (x^3 - 3x^2) =$

*Solution*: First, use Distributive Property:
$a(b + c) = ab + ac$.
Then: $-(x^3 - 3x^2) = -x^3 + 3x^2$. Therefore:
$(x^2 - 2x^3) - (x^3 - 3x^2) = x^2 - 2x^3 - x^3 + 3x^2$.
Now combine like terms:
$-2x^3 - x^3 = -3x^3$ and $x^2 + 3x^2 = 4x^2$.
Then:
$(x^2 - 2x^3) - (x^3 - 3x^2) = x^2 - 2x^3 - x^3 + 3x^2 = -3x^3 + 4x^2$.

**Example 2.** Add expressions. $(3x^3 - 5) + (4x^3 - 2x^2) =$

*Solution*: Remove parentheses:
$(3x^3 - 5) + (4x^3 - 2x^2) = 3x^3 - 5 + 4x^3 - 2x^2$.
Now, combine like terms:
$3x^3 - 5 + 4x^3 - 2x^2 = 7x^3 - 2x^2 - 5$.

**Example 3.** Simplify the expressions. $(-4x^2 - 2x^3) - (5x^2 + 2x^3) =$

*Solution*: First, use Distributive Property:
$-(5x^2 + 2x^3) = -5x^2 - 2x^3$.
Now, we have:
$(-4x^2 - 2x^3) - (5x^2 + 2x^3) = -4x^2 - 2x^3 - 5x^2 - 2x^3$.

Now, combine like terms and write in standard form:
$-4x^2 - 2x^3 - 5x^2 - 2x^3 = -4x^3 - 9x^2$.

# Multiplying and Dividing Monomials

- When you divide or multiply two monomials, you need to divide or multiply their coefficients and then divide or multiply their variables.

- In the case of exponents with the same base, for Division, subtract their powers, for Multiplication, add their powers.

- Exponent's Multiplication and Division rules:

$$x^a \times x^b = x^{a+b}, \frac{x^a}{x^b} = x^{a-b}$$

## Examples:

**Example 1.** Multiply expressions. $(3x^5)(9x^4) =$

*Solution*: Use multiplication property of exponents:
$x^a \times x^b = x^{a+b} \rightarrow x^5 \times x^4 = x^9$.
Then:
$(3x^5)(9x^4) = 27x^9$.

**Example 2.** Divide expressions. $\frac{12x^4y^6}{6xy^2} =$

*Solution*: Use division property of exponents:
$\frac{x^a}{x^b} = x^{a-b} \rightarrow \frac{x^4}{x} = x^{4-1} = x^3$
and
$\frac{y^6}{y^2} = y^{6-2} = y^4$.
Then:
$\frac{12x^4y^6}{6xy^2} = 2x^3y^4$.

**Example 3.** Divide expressions. $\frac{49a^6b^9}{7a^3b^4} =$

*Solution*: Use division property of exponents:
$\frac{x^a}{x^b} = x^{a-b} \rightarrow \frac{a^6}{a^3} = a^{6-3} = a^3$ and $\frac{b^9}{b^4} = b^{9-4} = b^5$.
Then:
$\frac{49a^6b^9}{7a^3b^4} = 7a^3b^5$.

bit.ly/3GmUpxZ

Find more at

# Multiplying a Polynomial and a Monomial

- When multiplying monomials, use the product rule for exponents.

$$x^a \times x^b = x^{a+b}$$

- When multiplying a monomial by a polynomial, use the distributive property.

$$a \times (b + c) = a \times b + a \times c = ab + ac$$
$$a \times (b - c) = a \times b - a \times c = ab - ac$$

## Examples:

**Example 1.** Multiply expressions. $6x(2x + 5)$

*Solution*: Use Distributive Property:

$a \times (b + c) = a \times b + a \times c = ab + ac$.

Therefore:
$6x(2x + 5) = 6x \times 2x + 6x \times 5$.
Now use the product rule for exponents:

$x^a \times x^b = x^{a+b}$.

Then:
$6x \times 2x + 6x \times 5 = 12x^2 + 30x$.

**Example 2.** Multiply expressions. $x(3x^2 + 4y^2)$

*Solution*: Use Distributive Property:
$x(3x^2 + 4y^2) = x \times 3x^2 + x \times 4y^2$.
And the product rule for exponents:
$x \times 3x^2 + x \times 4y^2 = 3x^3 + 4xy^2$.

**Example 3.** Multiply. $-x(-2x^2 + 4x + 5)$

*Solution*: Use Distributive Property:
$-x(-2x^2 + 4x + 5) = (-x)(-2x^2) + (-x) \times (4x) + (-x) \times (5)$.
Now simplify:
$(-x)(-2x^2) + (-x) \times (4x) + (-x) \times (5) = 2x^3 - 4x^2 - 5x$.

# Multiplying Binomials

- A binomial is a polynomial that is the sum or the difference of two terms, each of which is a monomial.

- To multiply two binomials, use the "FOIL" method. (First–Out–In–Last):

$$(x + a)(x + b) = x \times x + x \times b + a \times x + a \times b = x^2 + bx + ax + ab$$

## Examples:

**Example 1.** Multiply Binomials. $(x + 3)(x - 2) =$

*Solution*: Use "FOIL". (First–Out–In–Last):
$(x + 3)(x - 2) = x^2 - 2x + 3x - 6.$
Then combine like terms:
$x^2 - 2x + 3x - 6 = x^2 + x - 6.$

**Example 2.** Multiply. $(x + 6)(x + 4) =$

*Solution*: Use "FOIL". (First–Out–In–Last):
$(x + 6)(x + 4) = x^2 + 4x + 6x + 24.$
Then simplify:
$x^2 + 4x + 6x + 24 = x^2 + 10x + 24.$

**Example 3.** Multiply. $(x + 5)(x - 7) =$

*Solution*: Use "FOIL". (First–Out–In–Last):
$(x + 5)(x - 7) = x^2 - 7x + 5x - 35.$
Then simplify:
$x^2 - 7x + 5x - 35 = x^2 - 2x - 35.$

**Example 4.** Multiply Binomials. $(x - 9)(x - 5) =$

*Solution*: Use "FOIL". (First–Out–In–Last):
$(x - 9)(x - 5) = x^2 - 5x - 9x + 45.$
Then combine like terms:
$x^2 - 5x - 9x + 45 = x^2 - 14x + 45.$

bit.ly/3aCsOFL

Find more at

# Factoring Trinomials

To factor trinomials, you can use the following methods:

- "FOIL": $(x + a)(x + b) = x^2 + (b + a)x + ab$.

- "Difference of Squares":

$$a^2 - b^2 = (a + b)(a - b)$$
$$a^2 + 2ab + b^2 = (a + b)(a + b)$$
$$a^2 - 2ab + b^2 = (a - b)(a - b)$$

- "Reverse FOIL": $x^2 + (b + a)x + ab = (x + a)(x + b)$.

## Examples:

**Example 1.** Factor this trinomial. $x^2 - 2x - 8$

*Solution*: Break the expression into groups. You need to find two numbers that their product is $-8$ and their sum is $-2$.
(Remember "Reverse FOIL": $x^2 + (b + a)x + ab = (x + a)(x + b)$).
Those two numbers are 2 and $-4$. Then:
$x^2 - 2x - 8 = (x^2 + 2x) + (-4x - 8)$.
Now factor out $x$ from $x^2 + 2x$: $x(x + 2)$, and factor out $-4$ from $-4x - 8$: $-4(x + 2)$;
Then:
$(x^2 + 2x) + (-4x - 8) = x(x + 2) - 4(x + 2)$
Now factor out like term: $(x + 2)$. Then:
$(x + 2)(x - 4)$.

**Example 2.** Factor this trinomial. $x^2 - 2x - 24$

*Solution*: Break the expression into groups:
$(x^2 + 4x) + (-6x - 24)$.
Now factor out $x$ from $x^2 + 4x$: $x(x + 4)$, and factor out $-6$ from $-6x - 24$: $-6(x + 4)$;
Then:
$x(x + 4) - 6(x + 4)$,
 Now factor out like term:

$(x + 4) \rightarrow x(x + 4) - 6(x + 4) = (x + 4)(x - 6)$.

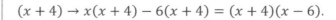

# Choosing a Factoring Method for Polynomials

- Factoring is the method to find the numbers or parts of a polynomial that you will use the product them to make a given number or smaller polynomials. Factoring is an important method for solving basic algebra problems; the skill of factorization becomes a necessary ability when working with different forms of polynomials. Factorization is a useful way to make it easy to deal with algebraic expressions and solve them.

- For solving polynomial equations, you can use the following factoring methods:
  - In a polynomial, regardless of the number of terms it has, always try to find the greatest common factor or GCF first. The GCF is the largest expression that is common in all of the terms. Using the greatest common factor is like doing the inverse of the distributive property.

  - If you have a trinomial expression with three terms, you can use the FOIL technique for factoring. The FOIL is a method to multiply binomials. The letters of FOIL imply First, Outside, Inside, and Last, and show the order of multiplying terms. For finding your answer you use multiplication for the first terms, then multiply outside terms, then multiply inside terms, then use multiplication for the last terms, and in the last step, you can combine like terms.

  - If you have a binomial expression with two terms, you should find the squares' difference, the cubes' difference, or the cubes' sum.

## Examples:

**Example 1.** Factor the polynomial expression completely. $3xy^2 - 21xy + 30x$

**Solution**: Try to find the GCF of all terms. The greatest common factor is $3x$: $3xy^2 - 21xy + 30x \rightarrow 3x(y^2 - 7y + 10)$. Here $b = -7$ and $c = 10$. So, find factors of 10 whose sum can be $-7$: $3x(y + \_)(y + \_) \rightarrow$ Therefore, the desired factors are $-2$ and $-5$: $3x(y - 2)(y - 5)$.

| factors of 10 | Sum |
|---|---|
| $-2$ and $-5$ | $-7$ |

**Example 2.** Factor the polynomial expression completely. $4xy^2 - 48xy + 144x$

**Solution**: Try to find the GCF of all terms. The greatest common factor is $4x$: $4xy^2 - 48xy + 144x \rightarrow 4x(y^2 - 12y + 36)$. Here $b = -12$ and $c = 36$. $y^2 - 12y + 36$ is a perfect square of the form $a^2 - 2ab + b^2$. Here $a = y$ and $b = 6$ so, $4x(y - 6)^2$.

bit.ly/3Xeore0

Find more at

# Factoring by GCF

- Factors are the numbers multiplied together in a multiplication problem. The solution to a multiplication problem is named product. In factoring a polynomial, you want to find the numbers that when multiplied create the polynomial.

- The greatest common factor or GCF of a polynomial is the greatest monomial that divides the polynomial's terms. Note that the GCF should be a factor of all terms in the polynomial.

- To find the GCF factor of a polynomial, you can take the following steps:
  - 1st step: Check every term in the polynomial. Find the GCF of all of the terms.
  - 2nd step: Now you have two factors: the GCF and another factor. Represent each term as a product of these two factors.
  - 3rd step: In the last step, to factor out the GCF you can use the distributive property.

## Examples:

**Example 1.** Factor the polynomial. $14x^4 + 2x^3$

*Solution*: First, check every term in the polynomial. Find the GCF of all of the terms: $14x^4 = 2 \times 7 \times x \times x \times x \times x$, $2x^3 = 2 \times x \times x \times x$. The common factors are $2 \times x \times x \times x$. So, the GCF is $2x^3$. Represent each term as a product of these two factors: $2x^3(7x) + 2x^3(1)$. In the last step, to factor out the GCF, you can use the distributive property: $2x^3(7x + 1)$.

**Example 2.** Factor the polynomial. $9y^4 + 12y^3 - 21y^2$

*Solution*: First find the GCF of all of the terms: $9y^4 = 3 \times 3 \times y \times y \times y \times y$, $12y^3 = 2 \times 2 \times 3 \times y \times y \times y$, $21y^2 = 3 \times 7 \times y \times y$. The common factors are $3 \times y \times y$. So, the GCF is $3y^2$. Represent each term as a product of these two factors: $3y^2(3y^2) + 3y^2(4y) - 3y^2(7)$. In the last step, to factor out the GCF, you can use the distributive property:

$$3y^2(3y^2 + 4y - 7) = 3y^2(y - 1)(3y + 7).$$

# Factors and Greatest Common Factors

- The Greatest Common Factor or GCF of numbers is the largest value of the number that you can find in the group of the common factors of the given numbers. If you have 2 natural numbers $x$ and $y$, GCF is the greatest realizable number that can divide both $x$ and $y$ with no remainder. To find GCF, these three methods can be used: Finding GCF using listing factors, Finding GCF using the prime factorization method, and Finding GCF using the division method.

- Finding GCF using listing factors: In this technique, you should list the factors of both numbers. Then you can easily check the common factors of the two numbers. Mark the common factors, and you can find the largest value amongst all of the common factors.

- Finding GCF using the prime factorization method: In this method, first, you can make a prime factored form and represent the numbers in this form. start from the least prime factor of each number. Then look at the factors that are common to each of the given numbers. GCF is the product of common prime factors of the given numbers.

- Finding GCF using the division method: In the division method, you create equal groups of objects or numbers. For the division of great numbers, you can use long division, because this method makes division problems easier. In finding GCF using the division method, GCF of numbers is the greatest positive integer that can divide all the given numbers, with no remainder.

## Examples:

**Example 1.** Find the GCF of 18, 24, and 42 by listing factors.

*Solution*: You should find each number's factors: Factors of 18 = {1,2,3,6,9,18}. Factors of 24 = {1,2,3,4,6,8,12,24}. Factors of 42 = {1,2,3,6,7,14,21,42}. Here following numbers are common factors between 18, 24, and 42: 1, 2, 3 and 6. 6 is the largest number so, GCF(18,24,42) = 6.

**Example 2.** Find the GCF of $5x^2$, $10x^4$, and $15x$.

*Solution*: You should find each number's factors: Factors of $5x^2 = 5 \times x \times x$. Factors of $10x^4 = 2 \times 5 \times x \times x \times x \times x$. Factors of $15x = 3 \times 5 \times x$. The following numbers are common factors between $5x^2$, $10x^4$, and $15x$: 5, and $x$. So, GCF$(5x^2, 10x^4, 15x) = 5x$.

bit.ly/3ZfrPHn
Find more at

# Operations with Polynomials

- The distributive property (Or the distributive property of multiplication over addition and subtraction) simplifies and solves expressions in the form of: $a(b + c)$ or $a(b - c)$.

- The distributive property is multiplying a term outside the parentheses by the terms inside.

- Distributive Property rule: $a(b + c) = ab + ac$.

## Examples:

**Example 1.** Simply using the distributive property. $(-2)(x + 3)$

*Solution*: Use the Distributive Property rule:
$a(b + c) = ab + ac$.
Then:
$(-2)(x + 3) = (-2 \times x) + (-2) \times (3) = -2x - 6$.

**Example 2.** Simply. $(-5)(-2x - 6)$

*Solution*: Use the Distributive Property rule:
$a(b + c) = ab + ac$.
Therefore:
$(-5)(-2x - 6) = (-5 \times -2x) + (-5) \times (-6) = 10x + 30$.

**Example 3.** Simply. $(7)(2x - 8) - 12x$

*Solution*: First, simplify $(7)(2x - 8)$ using the distributive property.
Then:
$(7)(2x - 8) = 14x - 56$.
Now, combine like terms:
$(7)(2x - 8) - 12x = 14x - 56 - 12x$.
In this expression, $14x$ and $-12x$ are "like" terms and we can combine them:
$14x - 12x = 2x$.
Then:
$14x - 56 - 12x = 2x - 56$.

# Even and Odd Functions

- A function $f$, is called even if $f(-x) = f(x)$ for all values of $x$ in the domain of the function.

- A function $f$, is called odd if $f(-x) = -f(x)$ for all values of $x$ in the domain of the function.

- An even function is symmetric relative to the $x$ −axis and an odd function is relative to the coordinate center.

## Examples:

**Example 1.** Identify whether the following function is even, odd, or neither:
$$f(x) = -2x^3$$

**Solution:** For this, it is enough to put $-x$ in the equation of the function and simplify:

$$f(-x) = -2(-x)^3 = -2(-x^3) = 2x^3 \rightarrow f(-x) = -f(x).$$

This means that the function is odd.

**Example 2.** Identify whether the following function is even, odd, or neither:
$$g(x) = x^4 - 4x^2 + 1$$

**Solution:** Put $-x$ in the equation of $g(x)$:

$$g(-x) = (-x)^4 - 4(-x)^2 + 1 = x^4 - 4x^2 + 1.$$

Since $g(-x) = g(x)$, therefore, this function is even.

**Example 3.** According to the graph, determine whether the function is even, odd, or neither.

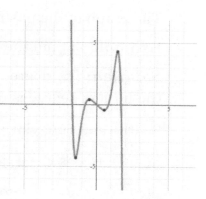

**Solution:** Considering that the graph is symmetric relative to the coordinate center (or in other words, relative to the $x$ −axis and the $y$ −axis), therefore the function corresponding to this graph is odd.

# End Behavior of Polynomial Functions

- Let $P$ be a polynomial function in the general form:

$$P(x) = a_n x^n + a_{n-1} x^{n-1} + \cdots + a_1 x + a_0.$$

- The domain of $P(x)$ is all real numbers, and is continuous for any value of $x$.

- The end behavior of $P(x)$ is according to the following table:

| End behavior of polynomial function $P(x) = a_n x^n + \cdots$ | | |
|---|---|---|
| If | $a_n > 0$ | $a_n < 0$ |
| $n$ is even | $P(x) \to +\infty$, as $x \to +\infty$ <br> $P(x) \to +\infty$, as $x \to -\infty$ | $P(x) \to -\infty$, as $x \to +\infty$ <br> $P(x) \to -\infty$, as $x \to -\infty$ |
| $n$ is odd | $P(x) \to +\infty$, as $x \to +\infty$ <br> $P(x) \to -\infty$, as $x \to -\infty$ | $P(x) \to -\infty$, as $x \to +\infty$ <br> $P(x) \to +\infty$, as $x \to -\infty$ |

## Examples:

**Example 1.** Find the end behavior of the function $f(x) = -x^4 + 3x^3 - x$.

*Solution:* The degree of the function is even, and the coefficient of the term with the largest degree is negative. Thus, the end behavior is:

$f(x) \to -\infty$, as $x \to +\infty$

$f(x) \to -\infty$, as $x \to -\infty$

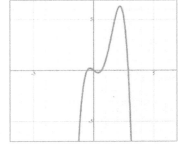

The graph of the polynomial function $f(x)$ is as follows.

**Example 2.** Find the end behavior of the function $g(x) = 2x^5 - 3x^3 + x^2 - 2$.

*Solution:* The degree of the function is odd, and the coefficient of the term with the largest degree is positive.

So, the end behavior is:

$g(x) \to +\infty$, as $x \to +\infty$

$g(x) \to -\infty$, as $x \to -\infty$

The graph of the polynomial function $g(x)$ is as follows.

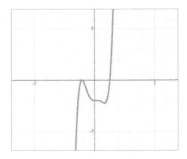

# Remainder and Factor Theorems

For a polynomial expression $P(x)$:

- Remainder theorem: The remainder of dividing $P(x)$ by $x - a$ is equal to $P(a)$.

- Factor theorem: $x - c$ is a factor of $P(x)$ if and only if $P(c) = 0$.

## Examples:

**Example 1.** Using the remainder theorem, find the remainder of dividing
$$P(x) = 3x^4 - 2x^3 + 1 \text{ by } x - 2.$$

*Solution:* According to the remainder theorem, calculate the value of the polynomial expression for 2. Therefore:

$$P(2) = 3(2)^4 - 2(2)^3 + 1 = 33.$$

The remainder of the division is equal to 33.

**Example 2.** Using the factor theorem, show that $x + 3$ is a factor of

$$x^3 - 6x + 9.$$

*Solution:* Let $Q(x) = x^3 - 6x + 9$. By using the factor theorem, since we want to check if $x + 3$ is a polynomial factor, we must put $x + 3 = 0$ and calculate the value of $Q(x)$ for $-3$. Therefore:

$$Q(-3) = (-3)^3 - 6(-3) + 9 = -27 + 18 + 9 = 0.$$

So, $x + 3$ is a factor of $x^3 - 6x + 9$.

**Example 3.** If $f(x)$ is a polynomial of degree 3, that leaves a remainder of zero when divided by $x$, $x + 1$ and $3x - 2$. What is the polynomial $f(x)$?

*Solution:* Since the remainder of the division by $x$, $x + 1$ and $x - 2$ is zero, according to the factor theorem, these polynomials are factors of $f(x)$. that means:

$$f(x) = x(x + 1)(x - 2)Q(x) \rightarrow f(x) = (x^3 - x^2 - 2x)Q(x).$$

$f(x)$ is a polynomial of degree 3, so, $Q(x)$ is equal to 1 (degree zero).

Therefore: $f(x) = x^3 - x^2 - 2x$.

# Polynomial Division (Long Division)

- Long division can be used to divide a polynomial $p(x)$ by the divisor in the form of $q(x)$.

- The long division of polynomials is similar to the long division of real numbers.

$$q(x) \overline{\big)\, p(x) = a_n x^n + a_{n-1} x^{n-1} + \cdots + a_0}^{\displaystyle Q(x)}$$

$$R(x)$$

Where $p(x)$, $q(x)$, $Q(x)$ and $R(x)$ are polynomials and the degree of $q(x)$ is less than $p(x)$:

- ○ $p(x)$ is the dividend.

- ○ $q(x)$ is the divisor.

- ○ $Q(x)$ is the quotient.

- ○ $R(x)$ is the remainder.

- The division continues until the degree of $R(x)$ is less than $q(x)$.

## Example:

Dividing polynomials: $(x^2 - 4x + 3) \div (x - 1)$.
**Solution:** Here, $(x - 1)$ is the divisor. Divide the first term of the dividend by the first term of the divisor:

$$x^2 \div x = x.$$

Write $x$ in the quotient and use it to multiply the divisor: $x(x - 1)$, and subtract the result from $(x^2 - 4x + 3)$. So:

$(x^2 - 4x + 3) - (x^2 - x) = x^2 - 4x + 3 - x^2 + x = -3x + 3.$

$$
\begin{array}{r}
x - 3 \\
x - 1 \,\overline{\big)\, x^2 - 4x + 3} \\
-x^2 + x \\
\hline
-3x + 3 \\
3x - 3 \\
\hline
0
\end{array}
$$

Now, divide the first term of the $(-3x + 3)$, by the first term of the divisor: $-3x \div x = -3$. Write $x$ in the quotient and use it to multiply the divisor: $-3(x - 1) = 3x - 3$ and subtract the last term of the division. It means: $-3x + 3 - (3x - 3) = 0$.

The remainder becomes zero.

# Polynomial Division (Synthetic Division)

- A method for dividing polynomials is a shortened form of long division.
- In synthetic division, you do not write the variables.
- To divide a polynomial $P(x) = a_n x^n + a_{n-1} x^{n-1} + \cdots + a_0$ by the divisor in the form $x - a$:

  Step1: First, write the polynomial coefficients of dividends in a row. Then write the constant term of the divisor on the left. That means $a$.

  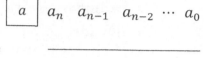

  Step2: Move the leading coefficient of the dividend to the bottom of the line.

  Step3: Multiply $a$ by $a_n$ and write the product $a \times a_n$ in the middle row.

  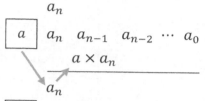

  Step4: Add $a_{n-1}$ and $a \times a_n$ in the second column and write the sum $a_k = a_{n-1} + a \times a_n$ in the bottom row.

  Step5: Continue steps 3 to 4 until the end of the division.

  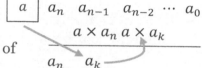

## Example:

Find the quotient:

$$(x^2 - 5x + 6) \div (x - 3).$$

**Solution:** Step1: First, write the polynomial coefficients of dividends in a row. Then write the constant term of the divisor on the left.

Step2: Move the leading coefficient of the dividend to the bottom of the line.

Step3: Multiply 3 by 1 and write product 3 in the middle row.

Step4: Add −5 and 3 in the second column and write the sum −2 in the bottom row.

Step5: Multiply 3 by −2 and write the product −6 in the middle row.

Step6: Add 6 and −6 in the second column and write the sum 0 in the bottom row. The remainder is 0, and $x - 2$ is the quotient.

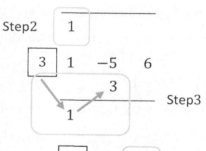

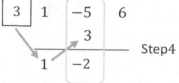

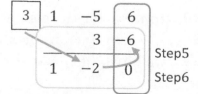

# Finding Zeros of Polynomials

- The zeros of a polynomial are values where $P(x) = 0$.

- In order to find the zeros of polynomials of different degrees, the following methods can be used:

  - A linear equation in form $mx + b = 0$: $x = -\frac{b}{m}$.
  - Quadratic equation: For a quadratic equation of the form $ax^2 + bx + c = 0$, using the formula:
  $$x = \frac{-b \pm \sqrt{b^2 - 2ac}}{2a}$$

  - Polynomials of degrees higher than 3 in the form $a_n x^n + a_{n-1} x^{n-1} + \cdots + a_1 x + a_0$ can be factored using the remainder theorem.
  - Using polynomial factorization by polynomial identities.

- If $P(x)$ is a polynomial and $P(a) > 0$ and $P(b) < 0$, then $P(x)$ has at least one real root between $a$ and $b$.

- The number of zeros of a polynomial is less than or equal to the degree of the polynomial.

## Examples:

**Example 1.** Find the zeros of the polynomial: $P(x) = x^3 - 3x^2 - 4x + 12$.

*Solution:* First, we factor the polynomial in such a way that it becomes the product of the first factors:

$x^3 - 3x^2 - 4x + 12 = x^2(x - 3) - 4(x - 3) = (x - 3)(x^2 - 4)$.

Now, for $x^2 - 4$ by factoring: $x^2 - 4 = (x - 2)(x + 2)$.

To find the zeros of the polynomial, solve the following equation: $x^3 - 3x^2 - 4x + 12 = 0$. Then: $x^3 - 3x^2 - 4x + 12 = 0 \rightarrow (x - 3)(x - 2)(x + 2) = 0$. So, $x - 3 = 0$, or $x - 2 = 0$, or $x + 2 = 0$. Therefore, the zeros of polynomial $P(x)$ are 3, 2 and $-2$.

**Example 2.** Find the zeros of the polynomial: $Q(x) = x^3 - 2x^2 - 3x$.

*Solution:* Factor the common factor of $x$: $x^3 - 2x^2 - 3x = x(x^2 - 2x - 3)$. Now, for $x^2 - 2x - 3$ by factoring: $x^2 - 2x - 3 = (x + 1)(x - 3)$. Then, to find the zeros of $Q(x)$, we have: $Q(x) = 0 \rightarrow x^3 - 2x^2 - 3x = x(x^2 - 2x - 3) = x(x + 1)(x - 3) = 0$.

Therefore, $x = 0$, $x + 1 = 0 \rightarrow x = -1$, and $x - 3 = 0 \rightarrow x = 3$ are the zeros of $Q(x)$.

# Polynomial identities

- In order to factorize or expand polynomials, we use polynomial identities.
- The most important polynomial identities are:
  - $(x + y)^2 = x^2 + 2xy + y^2$
  - $(x - y)^2 = x^2 - 2xy + y^2$
  - $(x + y)(x - y) = x^2 - y^2$
  - $(x + a)(x + b) = x^2 + (a + b)x + ab$
  - $(x + y)^3 = x^3 + 3x^2y + 3xy^2 + y^3 = x^3 + y^3 + 3xy(x + y)$
  - $(x - y)^3 = x^3 - 3x^2y + 3xy^2 - y^3 = x^3 - y^3 - 3xy(x - y)$
  - $(x + y + z)^2 = x^2 + y^2 + z^2 + 2xy + 2yz + 2zx$
  - $x^3 + y^3 = (x + y)(x^2 - xy + y^2)$
  - $x^3 - y^3 = (x - y)(x^2 + xy + y^2)$
  - $x^3 + y^3 + z^3 - 3xyz = (x + y + z)(x^2 + y^2 + z^2 - xy - yz - zx)$

- If $x + y + z = 0$, then $x^3 + y^3 + z^3 = 3xyz$.

## Examples:

**Example 1.** Factorize the following expressions: $9x^2 + y^2 + 6xy + 2y + 6x + 1$.

*Solution:* Using this identity: $(x + y + z)^2 = x^2 + y^2 + z^2 + 2xy + 2yz + 2zx$.

We have:

$9x^2 + y^2 + 6xy + 2y + 6x + 1 = (3x)^2 + y^2 + (1)^2 + 6xy + 2y(1) + 6(1)x$.

Then: $9x^2 + y^2 + 6xy + 2y + 6x + 1 = (3x + y + 1)^2$.

**Example 2.** Expand the following expressions: $(1 - x)^3$.

*Solution:* Using this identity:

$(x - y)^3 = x^3 - 3x^2y + 3xy^2 - y^3 = x^3 - y^3 - 3xy(x - y)$.

We have: $(1 - x)^3 = (1)^3 - 3(1)^2x + 3(1)x^2 - x^3$.

Thus: $(1 - x)^3 = 1 - 3x + 3x^2 - x^3$.

**Example 3.** Expand the following expressions: $(a - 2)(a^2 + 2a + 4)$.

*Solution:* Using this identity: $x^3 - y^3 = (x - y)(x^2 + xy + y^2)$.

Then: $(a - 2)(a^2 + 2a + 4) = a^3 - (2)^3$.

Therefore: $(a - 2)(a^2 + 2a + 4) = a^3 - 8$.

# Chapter 6: Practices

## ✎ Write each polynomial in standard form.

1) $2x - 5x =$

2) $5 + 12x - 8x =$

3) $x^2 - 2x^3 + 1 =$

4) $2 + 2x^2 - 1 =$

5) $-x^2 + 4x - 2x^3 =$

6) $-2x^2 + 2x^3 + 12 =$

7) $18 - 5x + 9x^4 =$

8) $2x^2 + 13x - 2x^3 =$

## ✎ Simplify each expression.

9) $2(4x - 6) =$

10) $5(3x - 4) =$

11) $x(2x - 5) =$

12) $4(5x + 3) =$

13) $2x(6x - 2) =$

14) $x(3x + 8) =$

15) $(x - 2)(x + 4) =$

16) $(x + 3)(x + 2) =$

## ✎ Add or subtract expressions.

17) $(x^2 - x) + (3x^2 - 5x) =$

18) $(x^3 + 2x) - (3x^3 + 2) =$

19) $(2x^3 - 4) + (2x^3 - 2) =$

20) $(-x^2 - 2) + (2x^2 + 1) =$

21) $(4x^2 + 3) - (3 - 3x^2) =$

22) $(x^3 + 3x^2) - (x^3 - 8) =$

23) $(7x - 9) + (3x + 5) =$

24) $(x^4 - 2x) - (x - x^4) =$

Effortless
Math
Education

### ✒ Simplify each expression.

25) $(-2x^4) \times (-5x^3) =$

26) $8x^8 \times -2x^2 =$

27) $5xy^4 \times 2x^2 =$

28) $-2x^6y \times 8xy =$

29) $3x^5 \times (-4x^3y^5) =$

30) $9x^3y^2 \times 3x^2y =$

### ✒ Simplify each expression.

31) $(x^2y)(xy^2) =$

32) $(x^4y^2)(2x^5y) =$

33) $(-2x^2y)(4x^4y^3) =$

34) $(-3x^5y^2)(2x^2y^4) =$

35) $(-4x^5y^3)(-2x^3y^4) =$

36) $(6x^6y^5)(3x^3y^8) =$

37) $\frac{-2x^4y^3}{x^2y^2} =$

38) $\frac{8x^4y^7}{2x^3y^4} =$

### ✒ Find each product.

39) $-2x(5x + 2y) =$

40) $3x(2x - y) =$

41) $4x(x + 5y) =$

42) $-4x(6x - 3) =$

43) $x(-2x + 9y) =$

44) $2x(5x - 8y) =$

45) $x(2x + 4y - 3) =$

46) $2x(x^2 - 2y^2) =$

47) $-4x(2x + 4y) =$

48) $3(x^2 + 7y^2) =$

49) $4x(-x^2y + 2y) =$

50) $5(x^2 - 4xy + 6) =$

51) $(x - 2)(x + 4) =$

52) $(x + 5)(x - 2) =$

**Effortless Math Education**

53) $(x - 3)(x - 4) =$

55) $(x - 6)(x - 3) =$

54) $(x + 2)(x + 2) =$

56) $(x + 5)(x + 7) =$

### ✍ Factor each trinomial.

57) $x^2 + 3x - 10 =$

60) $x^2 - 7x + 12 =$

58) $x^2 - x - 6 =$

61) $x^2 - x - 20 =$

59) $x^2 + 8x + 15 =$

62) $x^2 + 11x + 18 =$

### ✍ Factor the polynomials expression completely.

63) $2xy^2 + 20xy + 50x$

67) $5xy^2 - 90xy + 405x$

64) $3xy^2 - 36xy + 108x$

68) $4xy^2 - 56xy + 196x$

65) $4xy^2 + 24xy + 36x$

69) $4xy^2 - 20xy + 25x$

66) $7xy^2 + 56xy + 112x$

70) $8xy^2 + 48xy + 72x$

### ✍ Factor the polynomial.

71) $6x^4 - 10x^3$

75) $10x^5 - 30x^4$

72) $6x^4 - 18x^3$

76) $2x^5 - 16x^4 - 18x^3$

73) $2x^2 + 6x^3 - 8x^2$

77) $6x^5 - 10x^4 + 22x^3$

74) $5x^4 + 7x^3 + 10x^2$

78) $12x^8 + 9x^6 + 39x^4$

**Effortless**
**Math**
**Education**

## ✎ Find the GCF.

79) 5, 10 and 40

80) 8, 14 and 64

81) 26, 32 and 58

82) $6x^4$, $12x^4$ and $24x^4$

83) $14x^3$, $22x^6$ and $28x^5$

84) $36x^5$, $48x^4$ and $62x^6$

85) $78x^3$, $54x^4$ and $92x^4$

86) $8x^3$, $24x^3$ and $36x^6$

## ✎ Find each product.

87) $2(3x + 2) =$

88) $-3(2x + 5) =$

89) $4(7x - 3) =$

90) $5(2x - 4) =$

91) $3x(2x - 7) =$

92) $x^2(3x + 4) =$

93) $x^3(x + 5) =$

94) $x^4(5x - 3) =$

## ✎ Identify whether the following functions is even, odd, or neither.

95) $f(x) = x^2 + 6$

96) $f(x) = x^3 - 4x$

97) $f(x) = 4x^4 + 2$

98) $f(x) = 2x^3 - 2x + 2$

99) $f(x) = x^4 - 4x^2 + 4$

100) $f(x) = 2x^3 + 5x$

**Effortless Math Education**

**Find the end behavior of the functions.**

101) $f(x) = x^3 - 4x + 2$

104) $f(x) = -x^2 + 8x$

102) $f(x) = x^2 - 6x + 12$

105) $f(x) = -x^5 + 4x^3 - 2x - 4$

103) $f(x) = x^5 - 4x^3 + 4x + 2$

106) $f(x) = x^3 + 10x^2 + 22x + 4$

**Find the remainder of dividing .**

107) $f(x) = 2x^3 - 4x^2 + 1$ by $x - 2$.

108) $g(x) = x^3 + 5x^2 + 10x + 12$ by $x + 2$.

109) $k(x) = x^4 + 2x^3 - 15x^2 + 5$ by $x - 3$.

110) $t(x) = 3x^3 - 2x^2 - 12x + 8$ by $x - 4$.

111) $f(x) = -x^4 - 3x^3 + 4x$ by $x + 3$.

112) $k(x) = 3x^3 + 7x^2 - 3x^2 - 6$ by $x - 2$.

**Evaluate.**

113) $(x^2 + 2x - 36) \div (x - 5)$

116) $(x^2 - 3x - 21) \div (x - 7)$

114) $(x^2 + x - 79) \div (x + 9)$

117) $(x^3 + x^2 - 36x + 42) \div (x + 7)$

115) $(x^2 - x - 29) \div (x - 6)$

118) $(x^3 + 13x^2 + 42x + 54) \div (x + 9)$

 **Evaluate.**

119) $(x^3 - 13x^2 + 25x + 50) \div (x - 10)$

120) $(x^3 - 11x^2 + 26x + 20) \div (x - 5)$

121) $(x^3 + 15x^2 + 47x - 38) \div (x + 6)$

122) $(x^3 - 3x^2 - 3x - 2) \div (x - 2)$

✎ **Find the zeros of the polynomials.**

123) $x^3 + x^2 - 8x - 6$

124) $x^3 - 3x^2 - 4x + 12$

125) $x^3 + 3x^2 - 10x$

126) $x^3 + 2x^2 - 5x - 6$

127) $x^3 - 4x^2 - 5x$

128) $x^4 - x^3 - 20x^2$

✎ **Factorize the following expressions.**

129) $16x^2 + 4y^2 + 16xy + 12y + 24x + 9$

130) $-64x^3 + 96x^2 - 48x + 8$

131) $27x^3 + 27x^2 + 9x + 1$

132) $4x^2 + 4xy + 12x + y^2 + 6y + 9$

133) $125x^3 + 150x^2 + 60x + 8$

134) $8x^3 + 96x^2 + 384x + 512$

**Effortless
Math
Education**

# Chapter 6: Answers

1) $-3x$

2) $4x + 5$

3) $-2x^3 + x^2 + 1$

4) $2x^2 + 1$

5) $-2x^3 - x^2 + 4x$

6) $2x^3 - 2x^2 + 12$

7) $9x^4 - 5x + 18$

8) $-2x^3 + 2x^2 + 13x$

9) $8x - 12$

10) $15x - 20$

11) $2x^2 - 5x$

12) $20x + 12$

13) $12x^2 - 4x$

14) $3x^2 + 8x$

15) $x^2 + 2x - 8$

16) $x^2 + 5x + 6$

17) $4x^2 - 6x$

18) $-2x^3 + 2x - 2$

19) $4x^3 - 6$

20) $x^2 - 1$

21) $7x^2$

22) $3x^2 + 8$

23) $10x - 4$

24) $2x^4 - 3x$

25) $10x^7$

26) $-16x^{10}$

27) $10x^3 y^4$

28) $-16x^7 y^2$

29) $-12x^8 y^5$

30) $27x^5 y^3$

31) $x^3 y^3$

32) $2x^9 y^3$

33) $-8x^6 y^4$

34) $-6x^7 y^6$

35) $8x^8 y^7$

36) $18x^9 y^{13}$

37) $-2x^2 y$

38) $4xy^3$

39) $-10x^2 - 4xy$

40) $6x^2 - 3xy$

41) $4x^2 + 20xy$

42) $-24x^2 + 12x$

43) $-2x^2 + 9xy$

44) $10x^2 - 16xy$

45) $2x^2 + 4xy - 3x$

46) $2x^3 - 4xy^2$

47) $-8x^2 - 16xy$

48) $3x^2 + 21y^2$

49) $-4x^3 y + 8xy$

50) $5x^2 - 20xy + 30$

51) $x^2 + 2x - 8$

52) $x^2 + 3x - 10$

53) $x^2 - 7x + 12$

54) $x^2 + 4x + 4$

55) $x^2 - 9x + 18$

56) $x^2 + 12x + 35$

57) $(x - 2)(x + 5)$

58) $(x + 2)(x - 3)$

59) $(x + 5)(x + 3)$

Effortless
Math
Education

60) $(x-3)(x-4)$

61) $(x-5)(x+4)$

62) $(x+2)(x+9)$

63) $2x(y+5)^2$

64) $3x(y-6)^2$

65) $4x(y+3)^2$

66) $7x(y+4)^2$

67) $5x(y-9)^2$

68) $4x(y-7)^2$

69) $x(2y-5)^2$

70) $8x(y+3)^2$

71) $2x^3(3x-5)$

72) $6x^3(x-3)$

73) $6x^2(x-1)$

74) $x^2(5x^2+7x+10)$

75) $10x^4(x-3)$

76) $2x^3(x+1)(x-9)$

77) $2x^3(3x^2-5x+11)$

78) $3x^4(4x^4+3x^2+13)$

79) $GCF = 5$

80) $GCF = 2$

81) $GCF = 2$

82) $GCF = 6x^4$

83) $GCF = 2x^3$

84) $GCF = 2x^4$

85) $GCF = 2x^3$

86) $GCF = 4x^3$

87) $6x+4$

88) $-6x-15$

89) $28x-12$

90) $10x-20$

91) $6x^2-21x$

92) $3x^3+4x^2$

93) $x^4+5x^3$

94) $5x^5-3x^4$

95) Even

96) Odd

97) Even

98) Neither

99) Even

100) Odd

101) $f(x) \to -\infty$, as $x \to -\infty$

$f(x) \to +\infty$, as $x \to +\infty$

102) $f(x) \to +\infty$, as $x \to -\infty$

$f(x) \to +\infty$, as $x \to +\infty$

103) $f(x) \to -\infty$, as $x \to -\infty$

$f(x) \to +\infty$, as $x \to +\infty$

104) $f(x) \to -\infty$, as $x \to -\infty$

$f(x) \to -\infty$, as $x \to +\infty$

105) $f(x) \to +\infty$, as $x \to -\infty$

$f(x) \to -\infty$, as $x \to +\infty$

106) $f(x) \to -\infty$, as $x \to -\infty$

$f(x) \to +\infty$, as $x \to +\infty$

**Effortless**
**Math**
**Education**

107) 1

108) 4

109) 5

110) 120

111) $-12$

112) 34

113) $x + 7 - \frac{1}{x-5}$

114) $x - 8 - \frac{7}{x+9}$

115) $x + 5 + \frac{1}{x-6}$

116) $x + 4 + \frac{7}{x-7}$

117) $x^2 - 6x + 6$

118) $x^2 + 4x + 6$

119) $x^2 - 3x - 5$

120) $x^2 - 6x - 4$

121) $x^2 + 9x - 7 + \frac{4}{x+6}$

122) $x^2 - x - 5 - \frac{12}{x-2}$

123) $-3, 1 + \sqrt{3}, 1 - \sqrt{3}$

124) $3, 2, -2$

125) $0, 2, -5$

126) $-3, -1, 2$

127) $5, -1, 0$

128) $0, -4, 5$

129) $(4x + 2y + 3)^2$

130) $(-4x + 2)^3$

131) $(3x + 1)^3$

132) $(2x + y + 3)^2$

133) $(5x + 2)^3$

134) $(2x + 8)^3$

# CHAPTER

# 7 Functions Operations

Math topics that you'll learn in this chapter:

☑ Function Notation
☑ Adding and Subtracting Functions
☑ Multiplying and Dividing Functions
☑ Composition of Functions
☑ Writing Functions
☑ Parent Functions
☑ Function Inverses
☑ Inverse Variation
☑ Graphing Functions
☑ Domain and Range of Function
☑ Piecewise Function
☑ Positive, Negative, Increasing and Decreasing Functions on Intervals

# Function Notation

- Functions are mathematical operations that assign unique outputs to given inputs.

- Function notation is the way a function is written. It is meant to be a precise way of giving information about the function without a rather lengthy written explanation.

- The most popular function notation is $f(x)$ which is read as "$f$ of $x$". Any letter can name a function. For example: $g(x)$, $h(x)$, etc.

- To evaluate a function, plug in the input (the given value or expression) for the function's variable (place holder, $x$).

## Examples:

**Example 1.** Evaluate: $f(x) = x + 6$, find $f(2)$.

*Solution*: Substitute $x$ with 2:
Then:
$f(x) = x + 6 \rightarrow f(2) = 2 + 6 \rightarrow f(2) = 8$.

**Example 2.** Evaluate: $w(x) = 3x - 1$, find $w(4)$.

*Solution*: Substitute $x$ with 4:
Then:
$w(x) = 3x - 1 \rightarrow w(4) = 3(4) - 1 = 12 - 1 = 11$.

**Example 3.** Evaluate: $f(x) = 2x^2 + 4$, find $f(-1)$.

*Solution*: Substitute $x$ with $-1$:
Then:
$f(x) = 2x^2 + 4 \rightarrow f(-1) = 2(-1)^2 + 4 \rightarrow f(-1) = 2 + 4 = 6$.

**Example 4.** Evaluate: $h(x) = 4x^2 - 9$, find $h(2a)$.

*Solution*: Substitute $x$ with $2a$:
Then:
$h(x) = 4x^2 - 9 \rightarrow h(2a) = 4(2a)^2 - 9 \rightarrow h(2a) = 4(4a^2) - 9 = 16a^2 - 9$.

# Adding and Subtracting Functions

- Just like we can add and subtract numbers and expressions, we can add or subtract two functions and simplify or evaluate them. The result is a new function.

- For two functions $f(x)$ and $g(x)$, we can create two new functions:

$$(f + g)(x) = f(x) + g(x) \text{ and } (f - g)(x) = f(x) - g(x)$$

## Examples:

**Example 1.** If $g(x) = 2x - 2$, $f(x) = x + 1$. Find: $(g + f)(x)$.

*Solution*: We know that:
$(g + f)(x) = g(x) + f(x)$.
Then:
$(g + f)(x) = (2x - 2) + (x + 1) = 2x - 2 + x + 1 = 3x - 1$.

**Example 2.** If $f(x) = 4x - 3$, $g(x) = 2x - 4$. Find: $(f - g)(x)$.

*Solution*: Considering that:
$(f - g)(x) = f(x) - g(x)$.
Then:
$(f - g)(x) = (4x - 3) - (2x - 4) = 4x - 3 - 2x + 4 = 2x + 1$.

**Example 3.** If $g(x) = x^2 + 2$, and $f(x) = x + 5$. Find: $(g + f)(x)$.

*Solution*: According to the:
$(g + f)(x) = g(x) + f(x)$.
Then:
$(g + f)(x) = (x^2 + 2) + (x + 5) = x^2 + x + 7$.

**Example 4.** If $f(x) = 5x^2 - 3$, and $g(x) = 3x + 6$. Find: $(f - g)(3)$.

*Solution*: Use this:
$(f - g)(x) = f(x) - g(x)$.
Then:
$(f - g)(x) = (5x^2 - 3) - (3x + 6) = 5x^2 - 3 - 3x - 6 = 5x^2 - 3x - 9$.
Substitute $x$ with 3: $(f - g)(3) = 5(3)^2 - 3(3) - 9 = 45 - 9 - 9 = 27$.

bit.ly/3hdeFVO
Find more at

# Multiplying and Dividing Functions

- Just like we can multiply and divide numbers and expressions, we can multiply and divide two functions and simplify or evaluate them.

- For two functions $f(x)$ and $g(x)$, we can create two new functions:

$$(f.g)(x) = f(x).g(x) \text{ and } \left(\frac{f}{g}\right)(x) = \frac{f(x)}{g(x)}$$

## Examples:

**Example 1.** If $g(x) = x + 3$, $f(x) = x + 4$. Find: $(g.f)(x)$.

*Solution*: According to:
$(f.g)(x) = f(x).g(x)$.
Therefore:
$(x + 3)(x + 4) = x^2 + 4x + 3x + 12 = x^2 + 7x + 12$.

**Example 2.** If $f(x) = x + 6$, and $h(x) = x - 9$. Find: $\left(\frac{f}{h}\right)(x)$.

*Solution*: Use this:
$\left(\frac{f}{g}\right)(x) = \frac{f(x)}{g(x)}$.
We have:
$\left(\frac{f}{h}\right)(x) = \frac{x+6}{x-9}$.

**Example 3.** If $g(x) = x + 7$, and $f(x) = x - 3$. Find: $(g.f)(2)$.

*Solution*: Considering that:
$(f.g)(x) = f(x).g(x)$.
We have:
$(g.f)(x) = (x + 7)(x - 3) = x^2 - 3x + 7x - 21 = x^2 + 4x - 21$.
Substitute $x$ with 2:
$(g.f)(x) = (2)^2 + 4(2) - 21 = 4 + 8 - 21 = -9$.

**Example 4.** If $f(x) = x + 3$, and $h(x) = 2x - 4$. Find: $\left(\frac{f}{h}\right)(3)$.

*Solution*: We have: $\left(\frac{f}{h}\right)(x) = \frac{f(x)}{h(x)} = \frac{x+3}{2x-4}$.

Substitute $x$ with 3:
$\left(\frac{f}{h}\right)(3) = \frac{3+3}{2(3)-4} = \frac{6}{2} = 3$.

# Composition of Functions

- "Composition of functions" simply means combining two or more functions in a way where the output from one function becomes the input for the next function.

  The notation used for composition is: $(fog)(x) = f(g(x))$ and is read "$f$ composed with $g$ of $x$" or "$f$ of $g$ of $x$".

## Examples:

**Example 1.** Using $f(x) = 2x + 3$ and $g(x) = 5x$, find: $(fog)(x)$.

*Solution*: Using definition:
$(fog)(x) = f(g(x))$.
Then:
$(fog)(x) = f(g(x)) = f(5x)$.
Now, find $f(5x)$ by substituting $x$ with $5x$ in $f(x)$ function.
Then:
$f(x) = 2x + 3; (x \rightarrow 5x) \rightarrow f(5x) = 2(5x) + 3 = 10x + 3$.

**Example 2.** Using $f(x) = 3x - 1$ and $g(x) = 2x - 2$, find: $(gof)(5)$.

*Solution*:
$(fog)(x) = f(g(x))$.
Then: Using by:
$(gof)(x) = g(f(x)) = g(3x - 1)$.
Now, substitute $x$ in $g(x)$ by $(3x - 1)$.
Then:
$g(3x - 1) = 2(3x - 1) - 2 = 6x - 2 - 2 = 6x - 4$.
Substitute $x$ with 5: $(gof)(5) = g(f(5)) = 6(5) - 4 = 30 - 4 = 26$.

**Example 3.** Using $f(x) = 2x^2 - 5$ and $g(x) = x + 3$, find: $f(g(3))$.

*Solution*: First, find $g(3)$: $g(x) = x + 3 \rightarrow g(3) = 3 + 3 = 6$.
Then: $f(g(3)) = f(6)$.
Now, find $f(6)$ by substituting $x$ with 6 in $f(x)$ function.
$f(g(3)) = f(6) = 2(6)^2 - 5 = 2(36) - 5 = 67$.

Find more at
bit.ly/2WHBkAg

# Writing Functions

- A function is a kind of relationship between variables and it shows how 2 things are related to each other. It can take various forms, like an equation.
- Every function consists of inputs and outputs. The input is a variable that goes into the function it is also called the independent variable or domain. The output is a variable that comes out of the function and it is also called the dependent variable or range.
- The function rule is an algebraic statement that specifies a function. The function determines which inputs are suitable, and the outputs will be determined by the inputs. 3 methods to show functions are graphs, tables, and algebraic expressions.
- A table that shows a function usually includes a column of inputs, a column of outputs, and a third column between the input and output to present how the outputs can be made by the inputs.

## Examples:

**Example 1.** According to the values of $x$ and $y$ in the following relationship, find the right equation. $\{(1,4),(2,8),(3,12),(4,16)\}$

*Solution*: Find the relationship between the first $x$ −value and first $y$ −value: $(1,4)$. The value of $y$ is 4 times the value of $x$: $1 \times 4 = 4$. The equation is $y = 4x$. Now draw the input-output table of the values to make sure the relationship is correct:

| $x$ | $y = 4x$ | $y$ |
|-----|----------|-----|
| 1 | $4(1) = 4$ | 4 |
| 2 | $4(2) = 8$ | 8 |
| 3 | $4(3) = 12$ | 12 |
| 4 | $4(4) = 16$ | 16 |

**Example 2.** According to the values of $x$ and $y$ in the following relationship, find the right equation. $\{(1,3),(2,4),(3,5),(4,6)\}$

*Solution*: Find the relationship between the first $x$ −value and first $y$ −value: $(1,3)$. The value of $y$ is 2 more than the $x$ −value: $1 + 2 = 3$. So, the equation is $y = x + 2$. Now check the other values: $2 + 2 = 4, 3 + 2 = 5, 4 + 2 = 6$. Therefore each $x$ −value and $y$ −value satisfies the equation $y = x + 2$.

# Parent Functions

- The parent function is the simplest form of representation of any function without transformations.

- The most important parent functions include constant, linear, absolute-value, polynomials, rational, radical, exponential and logarithmic functions.

- For the function $f(x)$ and constant number $k > 0$, the transformations of functions in which the properties of the parent function are preserved is as follows:

| $y = f(x) + k$ | $y = f(x - k)$ | $y = -f(x)$ |
|---|---|---|
| If $k > 0$, shifted to the up<br><br>If $k < 0$, shifted to the down | If $k > 0$, shifted to the right<br><br>If $k < 0$, shifted to the left | is symmetric to the function<br>$y = f(x)$ with respect to the $x$ −axis. |
| $y = kf(x)$ | $y = f(kx)$ | $y = f(-x)$ |
| If $k > 1$, $f(x)$ is stretch vertically<br><br>If $0 < k < 1$, $f(x)$ is compressed vertically | If $k > 1$, $f(x)$ is compressed horizontally<br><br>If $0 < k < 1$, $f(x)$ is stretch horizontally | is symmetric to the function<br>$y = f(x)$ with respect to the $y$ −axis. |

## Example:

What is the parent graph of the following function and what transformations have taken place on it: $y = 2x^2 - 1$.

**Solution:** We know that the graph $y = f(x) + k$; $k < 0$, is shifted $k$ units to the down of the graph $y = f(x)$. Then, $y = 2x^2 - 1$, is shifted 1 unit down of the graph $y = 2x^2$.

On the other hand, $y = kf(x)$; $k > 1$ is stretched vertically then the function $y = x^2$ is stretched vertically by a factor of 2.

Clearly, you can see that the function $y = x^2$ is the parent function of $y = 2x^2 - 1$.

bit.ly/3JdVbR2

Find more at

# Function Inverses

-   An inverse function is a function that reverses another function: If the function $f$ applied to an input $x$ gives a result of $y$, then applying its inverse function $g$ to $y$ gives the result $x$. $f(x) = y$ if and only if $g(y) = x$.

-   The inverse function of $f(x)$ is usually shown by $f^{-1}(x)$.

## Examples:

**Example 1.** Find the inverse of the function: $f(x) = 2x - 1$.

*Solution*: First, replace $f(x)$ with $y$:

$y = 2x - 1$.

Then, replace all x's with y and all y's with $x$:

$x = 2y - 1$.

Now, solve for y:

$x = 2y - 1 \rightarrow x + 1 = 2y \rightarrow \frac{1}{2}x + \frac{1}{2} = y$.

Finally replace $y$ with $f^{-1}(x)$:

$f^{-1}(x) = \frac{1}{2}x + \frac{1}{2}$.

**Example 2.** Find the inverse of the function: $g(x) = \frac{1}{5}x + 3$.

*Solution*: First, replace $g(x)$ with $y$:

$g(x) = \frac{1}{5}x + 3 \rightarrow y = \frac{1}{5}x + 3$,

Then, replace all x's with y and all y's with $x$:

$x = \frac{1}{5}y + 3$,

Now, solve for y:

$x - 3 = \frac{1}{5}y \rightarrow 5(x - 3) = y \rightarrow y = 5x - 15 \rightarrow g^{-1} = 5x - 15$.

**Example 3.** Find the inverse of the function: $h(x) = \sqrt{x} + 6$.

*Solution*: $h(x) = \sqrt{x} + 6 \rightarrow y = \sqrt{x} + 6$, replace all x's with y and all y's with $x$:

$\rightarrow x = \sqrt{y} + 6 \rightarrow x - 6 = \sqrt{y} \rightarrow (x - 6)^2 = (\sqrt{y})^2 \rightarrow x^2 - 12x + 36 = y$

$\rightarrow h^{-1}(x) = x^2 - 12x + 36$.

# Inverse Variation

- Inverse variation is a type of proportion in which one quantity decreases while another increases or vice versa. This means that the amount or absolute value of one quantity decreases if the other quantity increases so that their product always remains the same. This product is also known as the proportionality constant.

- An inverse variation is represented by the equation $xy = k$ or $y = \frac{k}{x}$. That is, if there is some non-zero constant $k$, $y$ changes inversely so that $xy = k$ or $y = \frac{k}{x}$ where $x \neq 0$, $y \neq 0$.

- If inverse variation includes the points $(x_1, y_1)$ and $(x_2, y_2)$, inverse variation can be represented by the equation $x_1 y_1 = k$ and $x_2 y_2 = k$.

## Examples:

**Example 1.** If $y$ varies inversely as $x$ and $y = 12$ when $x = 5$. What's the value of $y$ when $x$ is 3?

**Solution**: Use $y = 12$ and $x = 5$ to find the value of $k$: $xy = k \rightarrow k = 5 \times 12 \rightarrow k = 60$. Now that we have found $k$, we can again use the value of $k$, which is 60, and the value of $x$, which is 3, to get the value of $y$: $k = xy \rightarrow 60 = 3 \times y \rightarrow y = 20$.

**Example 2.** An inverse variation includes points $(4, 9)$ and $(12, m)$. Find $m$.

**Solution**: First find the variation's constant. Plug $x_1 = 4$ and $y_1 = 9$ in to the equation $x_1 y_1 = k$ and then solve for $k$: $x_1 y_1 = k \rightarrow k = 4 \times 9 = 36$. Now use the inverse variation equation $x_2 y_2 = k$ to find $m$ or $y_2$ when $x_2 = 12$:
$x_2 y_2 = k \rightarrow 12 \times m = 36 \rightarrow m = \frac{36}{12} = 3 \rightarrow m = 3$.

**Example 3.** If $y$ varies inversely as $x$ and $y = 0.5$ when $x = 8$. What's the value of $x$ when $y$ is 2?

**Solution**: Use $y = 0.5$ and $x = 8$ to find the value of $k$: $xy = k \rightarrow k = 8 \times 0.5 \rightarrow k = 4$. Now that we have found $k$ which is 4, and the value of $y$ which is 2 to get the value of $x$:
$k = xy \rightarrow 4 = x \times 2 \rightarrow x = \frac{4}{2} \rightarrow x = 2$.

bit.ly/3ilj6mt

Find more at

# Graphing Functions

- Graphing functions is a way to show the function on the coordinate plane and to achieve this purpose, you can draw a line or a curve that represents a given function on the coordinate plane. A graph represents a function when each point on the line (or curve) of the graph can be an answer to the function equation.

- To graph linear functions and quadratic functions, first determine the shape of the graph. If it's a linear function and it's in form $f(x) = ax + b$, then this graph should be a line. If it's a quadratic function and it's in form $f(x) = ax^2 + bx + c$, then this graph should be a parabola.

- To graph linear functions, consider some random points on it, then substitute some random $x-$values in the equation and find the related $y-$values of the equation, to be able to graph the ordered pairs.

- To graph a quadratic function, you can consider some random points on it. But the graph of a quadratic function is a perfect $U-$shape, so to find a perfect $U-$shaped parabola, you should find its vertex. This means finding the point that the curve is turning. The next step after finding the vertex is finding 2 or 3 random points on every side of the vertex and then graphing the parabola.

## Example:

Graph the following function. $y = 3x - 2$
*Solution*: First, consider some random $x-$values, make an input-output table, and find their ordered pairs. These ordered pairs satisfy the function. Now, use the ordered pairs and draw a line that passes through all the points.

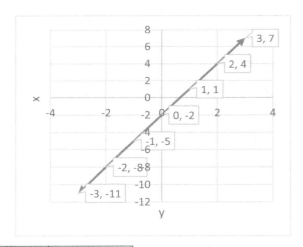

| $x$ | $y = 3x - 2$ | $(x, y)$ |
|-----|-------------|----------|
| $-3$ | $3(-3) - 2 = -11$ | $(-3, -11)$ |
| $-2$ | $3(-2) - 2 = -8$ | $(-2, -8)$ |
| $-1$ | $3(-1) - 2 = -5$ | $(-1, -5)$ |
| $0$ | $3(0) - 2 = -2$ | $(0, -2)$ |
| $1$ | $3(1) - 2 = 1$ | $(1, 1)$ |
| $2$ | $3(2) - 2 = 4$ | $(2, 4)$ |
| $3$ | $3(3) - 2 = 7$ | $(3, 7)$ |

# Domain and Range of Function

- For the function $f(x)$:

  ○ The set of all possible inputs is called the domain of the function.
  ○ The set of all possible outputs for each value of the domain is called the range of the function.

- The following methods can be used to obtain the domain of functions:

  ○ The domain of a polynomial function is all real numbers.
  ○ The domain of a square root function of the form $y = \sqrt{f(x)}$ is all values of $x$ where, $f(x) \geq 0$.
  ○ The domain of an exponential function is all real numbers.
  ○ The domain of a logarithmic function of the form $y = f(x)$ is all values of $x$ where, $f(x) > 0$.
  ○ Domain of a rational function of the form $y = \frac{f(x)}{g(x)}$, is all value of $x$ where, $g(x) \neq 0$.
  ○ The domain of a piecewise function is the union of all the smaller domains.

- The domain of a function on the graph is the image of the graph on the $x$ −axis.
- The range of a function on the graph is the image of the graph on the $y$ −axis.

## Examples:

**Example 1.** Find the domain and range of the function $f(x) = \frac{1}{x-1}$.

*Solution:* Since the domain of the rational function is a set of all real numbers except $x - 1 = 0$, then we have: $x - 1 = 0 \rightarrow x = 1$.

To find the range of the function, we know that, no matter how large or small $x$ becomes, $f(x)$ will never be equal to zero. So, the range of $f(x)$ is all real numbers except zero.

**Example 2.** Find the domain and range of the function $g(x) = \sqrt{1 - x} + 2$.

*Solution:* For domain, find non-negative values for radicals: $1 - x \geq 0$. Then the domain of the function is $1 - x \geq 0 \rightarrow x \leq 1$.

For range, we know that $\sqrt{1 - x} \geq 0$, so $\sqrt{1 - x} + 2 \geq 0 + 2 \rightarrow$ $\sqrt{1 - x} + 2 \geq 2$. Therefore: $g(x) \geq 2$.

bit.ly/3iim779

Find more at

# Piecewise Function

- A piecewise function uses more than one formula to define domain values.

- In general, a piecewise function can be shown as follows:

$$f(x) = \begin{cases} f_1(x), & x \in Domain\ f_1(x) \\ f_2(x), & x \in Domain\ f_2(x) \\ \vdots \\ f_n(x), & x \in Domain\ f_n(x) \end{cases}$$

Where $n \geq 2$, and the domain of the function is the union of all of the smaller domains.

- Absolute value functions can be shown as piecewise functions.

## Examples:

**Example 1.** Graph: $f(x) = \begin{cases} 2, & x < 0 \\ x + 1, & x \geq 1 \end{cases}$

***Solution:*** Step1: Graph $y = 2$ for $x < 0$. As follow:

Step2: Graph $y = x + 1$ for $x \geq 1$.

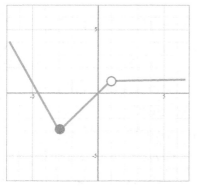

**Example 2.** Write the piecewise function represented by the graph.

***Solution:*** According to the graph, this function contains three pieces. The domain of the function is the union of three intervals as $(-\infty, -3]$, $[-3, 1)$, and $(1, +\infty)$. Using two points on the graph of each piece, we write the linear function from them.

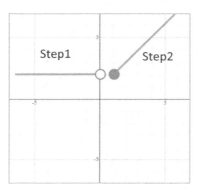

Step1: The interval $(-\infty, -3]$, two points $(-4, -1)$ and $(-3, -3)$ lie on the line $y = -2x - 9$.

Step2: The interval $[-3, 1)$, two points $(-3, -3)$ and $(1, 1)$ lie on the line $y = x$.

Step3: The interval $(1, +\infty)$, the graph is a horizontal function passing through the point $(1, 1)$.

Therefore: $f(x) = \begin{cases} -2x - 9 & x < -3 \\ x & -3 \leq x < 1. \\ 1 & x > 1 \end{cases}$

# Positive, Negative, Increasing and Decreasing Functions on Intervals

- For the function $f(x)$ over an arbitrary interval $I$:

  - $f$ is called positive if for every $x$ in the interval $I$, $f(x) > 0$. (Above the $x$ −axis)
  - $f$ is called negative if for every $x$ in the interval $I$, $f(x) < 0$. (Under the $x$ −axis)

- For the given function $f$, consider the points $x$ and $y$ in an arbitrary open interval of the domain. in this case:

  - The function $f$ is increasing if for every $x < y$ in the interval, $f(x) \leq f(y)$.
  - The function $f$ is decreasing if for every $x < y$ in the interval, $f(x) \geq f(y)$.
  - The function $f$ is constant if for every $x < y$ in the interval, $f(x) = f(y) = C$, such that $C$ is a constant number.

## Example:

According to the graph, determine in which intervals the function is positive, negative, increasing, or decreasing.

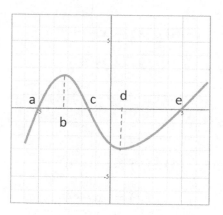

*Solution*: Considering the graph, you see that:

- For $(-\infty, a)$ and $(c, e)$, the graph is under the $x$ −axis. So, the function corresponding to the graph is negative.
- For $(a, c)$ and $(e, +\infty)$, the graph is above the $x$ −axis. So, the function is positive.
- Since the values of the function increase in the intervals $(-\infty, b)$ and $(d, +\infty)$, the corresponding function of the graph is increasing in each interval.
- In the interval $(b, d)$, the function is decreasing.

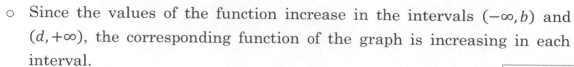

bit.ly/3GLpoFb

Find more at

# Chapter 7: Practices

## ✎ Evaluate each function.

1) $g(n) = 2n + 5$, find $g(2)$

———

2) $h(n) = 5n - 9$, find $h(4)$

———

3) $k(n) = 10 - 6n$, find $k(2)$

———

4) $g(n) = -5n + 6$, find $g(-2)$

———

5) $k(n) = -8n + 3$, find $k(-6)$

———

6) $w(n) = -2n - 9$, find $w(-5)$

———

## ✎ Perform the indicated operation.

7) $f(x) = x + 6$

$g(x) = 3x + 2$

Find $(f - g)(x)$

———

8) $g(x) = x - 9$

$f(x) = 2x - 1$

Find $(g - f)(x)$

———

9) $h(x) = 5x + 6$

$g(x) = 2x + 4$

Find $(h + g)(x)$

———

10) $g(x) = -6x + 1$

$f(x) = 3x^2 - 3$

Find $(g + f)(5)$

———

11) $g(x) = 7x - 1$

$h(x) = -4x^2 + 2$

Find $(g - h)(-3)$

———

12) $h(x) = -x^2 - 1$

$g(x) = -7x - 1$

Find $(h - g)(-5)$

———

**Effortless Math Education**

## ✏ Perform the indicated operation.

13) $g(x) = x + 3$

    $f(x) = x + 1$

    Find $(g. f)(x)$

    _____

14) $f(x) = 4x$

    $h(x) = x - 6$

    Find $(f. h)(x)$

    _____

15) $g(a) = a - 8$

    $h(a) = 4a - 2$

    Find $(g. h)(3)$

    _____

16) $f(x) = 6x + 2$

    $h(x) = 5x - 1$

    Find $\left(\frac{f}{h}\right)(-2)$

    _____

17) $f(x) = 7x - 1$

    $g(x) = -5 - 2x$

    Find $\left(\frac{f}{g}\right)(-4)$

    _____

18) $g(a) = a^2 - 4$

    $f(a) = a + 6$

    Find $\left(\frac{g}{f}\right)(-3)$

    _____

## ✏ Using $f(x) = 4x + 3$ and $g(x) = x - 7$, find:

19) $g\big(f(2)\big) =$ _____

20) $g\big(f(-2)\big) =$ _____

21) $f\big(g(4)\big) =$ _____

22) $f\big(f(7)\big) =$ _____

23) $g\big(f(5)\big) =$ _____

24) $g\big(f(-5)\big) =$ _____

## ✏ According to the values of $x$ and $y$ in the following relationship, find the right equation.

25) $\{(1,3), (2,6), (3,9), (4,12)\}$    _____

26) $\{(1,5), (2,7), (3,9), (4,11)\}$    _____

Effortless
Math
Education

What is the parent graph of the following function and what transformations have taken place on it?

27) $y = x^2 - 3$

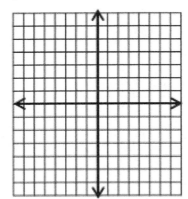

28) $y = x^3 + 4$

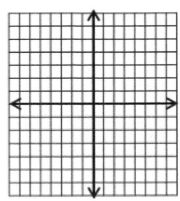

Find the inverse of each function.

29) $f(x) = \frac{1}{x} - 6 \rightarrow f^{-1}(x) =$

32) $h(x) = \frac{2x - 10}{4} \rightarrow h^{-1}(x) =$

30) $g(x) = \frac{7}{-x-3} \rightarrow g^{-1}(x) =$

33) $f(x) = \frac{-15+x}{3} \rightarrow f^{-1}(x) =$

31) $h(x) = \frac{x+9}{3} \rightarrow h^{-1}(x) =$

34) $s(x) = \sqrt{x} - 2 \rightarrow s^{-1}(x) =$

Solve.

35) If $y$ varies inversely as $x$ and $y = 18$ when $x = 6$. What's the value of $y$ when $x$ is 4? _____

36) If $y$ varies inversely as $x$ and $y = 0.8$ when $x = 6$. What's the value of $x$ when $y$ is 3? _____

**Effortless Math Education**

✎ **Graph the following functions.**

37) $y = 2x - 5$

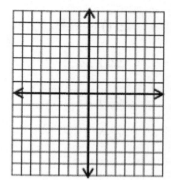

38) $y = 4x + 3$

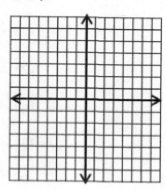

✎ **Find the domain and range of the functions.**

39) $y = x^3 - 4$

   Domain: _____

   Range: _____

40) $y = \sqrt{x - 8} + 4$

   Domain: _____
   Range: _____

41) $y = \frac{2}{2x-1}$

   Domain: _____
   Range: _____

42) $y = -2x^3 + 6$

   Domain: _____
   Range: _____

✎ **Graph.**

43) $f(x) = \begin{cases} 4 - 2, & x < -1 \\ x - 1, & x \geq 0 \end{cases}$

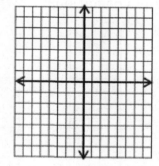

✎ **Solve.**

44) Determine the intervals where the function is increasing and decreasing.
    Submit your solution in interval notation.

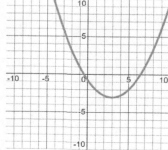

**Effortless**
**Math**
**Education**

# Chapter 7: Answers

1) 9

2) 11

3) −2

4) 16

5) 51

6) 1

7) −2x + 4

8) −x − 8

9) 7x + 10

10) 43

11) 12

12) −60

13) $x^2 + 4x + 3$

14) $4x^2 − 24x$

15) −50

16) $\frac{10}{11}$

17) $-\frac{29}{3}$

18) $\frac{5}{3}$

19) 4

20) −12

21) −9

22) 127

23) 16

24) −24

25) $y = 3x$

26) $y = 2x + 3$

27) Parent: Quadratic
Transformations: Down 3

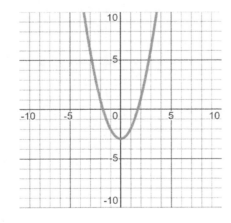

28) Parent: Cubic
Transformations: Up 4

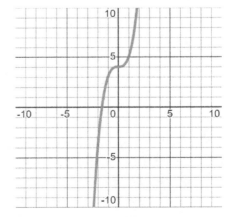

**Effortless
Math
Education**

29) $f^{-1}(x) = \frac{1}{x+6}$

30) $g^{-1}(x) = -\frac{7+3x}{x}$

31) $h^{-1}(x) = 3x - 9$

32) $h^{-1}(x) = 2x + 5$

33) $f^{-1}(x) = 3x + 15$

34) $s^{-1}(x) = x^2 + 4x + 4$

35) $y = 27$

36) $x = 1.6$

37)

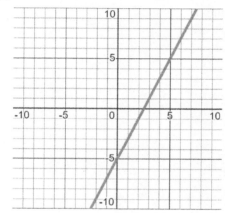

38)

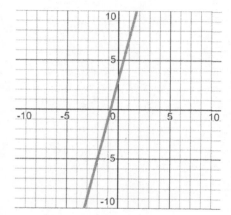

39) Domain: $-\infty < x < \infty$

Range: $-\infty < f(x) < \infty$

40) Domain: $x \geq 8$

Range: $f(x) \geq 4$

41) Domain: $x < \frac{1}{2}$ or $x > \frac{1}{2}$

Range: $f(x) < 0$ or $f(x) > 0$

42) Domain: $-\infty < x < \infty$

Range: $-\infty < f(x) < \infty$

43)

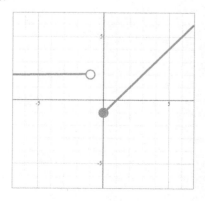

44) Increasing intervals: $(3, +\infty)$

Decreasing intervals: $(-\infty, 3)$

**Effortless**
**Math**
**Education**

# CHAPTER

# 8 Exponential Functions

Math topics that you'll learn in this chapter:

- ☑ Exponential Function
- ☑ Linear, Quadratic and Exponential Models
- ☑ Linear vs Exponential Growth

115

# Exponential Function

- An exponential function with a base $a$ is shown as $f(x) = a^x$, where $a$ is a positive real number and $a \neq 1$, and $a \in \mathbb{R}$.

- Exponential Growth and Decay:
  - If $a > 1$, the function $f(x) = a^x$ is exponential growth.
  - If $0 < a < 1$, the function $f(x) = a^x$ is exponential decay.
- All exponent functions of the form $f(x) = a^x$ have the same domain, range and $y$ −intercept. As follow: Domain: $\mathbb{R}$, Range: $(0, +\infty)$, $y$ −intercept: 1

## Example:

Graph $f(x) = 2^x$, and $g(x) = -5f(x) + 1$. Which one is exponential decay? Then give domain, range and $y$ −intercept.

**Solution:** Since the base of the $f(x) = 2^x$ is 2 and greater than 1, then $f(x)$ is growth with domain $\mathbb{R}$, range $(0, +\infty)$, and $y$ −intercept 1.

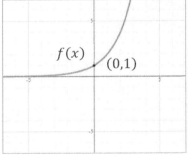

We know that the function $y = h(x)$ is symmetric to the function $y = -kf(x)$ with respect to the $x$ −axis, and all values of function $y = kf(x)$, where $k > 0$ are $k$ times of function $y = h(x)$. Now, all values of $y = 5f(x)$ are 5 times the function $y = f(x)$, and the function $y = 5f(x)$ is symmetric to the function $y = -5f(x)$ with respect to the $x$ −axis. As follow:

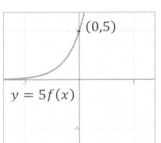

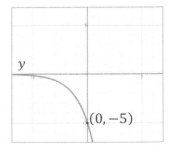

On the other hand, $y = -5f(x) + 1$ is shifted 1 unit to the up, therefore, according to the graph of $y = -5f(x) + 1$, the $y$ −intercept is $-4$, the domain is $\mathbb{R}$, and the range is $(-\infty, 1)$.

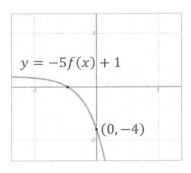

# Linear, Quadratic and Exponential Models

- Functions are a kind of mathematical relationship between inputs and outputs and linear, quadratic, and exponential functions are the 3 main types of functions.
- A linear function is a type of function whose highest exponent is 1 and the standard form of this function is $y = mx + b$. The graph of the linear function shows a straight line; therefore, its name is a linear function.
- Quadratic functions when graphed make parabolas. A quadratic function is a type of polynomial function whose highest exponent is 2. The quadratic function's standard form is $y = ax^2 + bx + c$.

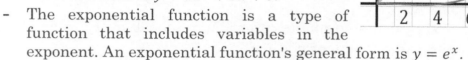

- The exponential function is a type of function that includes variables in the exponent. An exponential function's general form is $y = e^x$.
- The graphs of linear functions are straight lines with no curve. The graphs of quadratic functions are parabola-shaped. The graphs of exponential function have a curve. This curve can be vertical in the beginning and then grow to be horizontal or can be horizontal in the beginning and then become more vertical.
- You'll be able to find the degree of the model for a given ordered pair's information by determining the differences between dependent values:
- The model is linear if the first difference is constant.
- The model is quadratic if the second difference is constant.
- The model is exponential if the independent variable changes by a constant ratio.

## Example:

Determine whether the following table of values represents a linear function, an exponential function, or a quadratic function.

| $x$ | $-5$ | $-3$ | $-1$ | $1$ | $3$ |
|-----|------|------|------|-----|-----|
| $y$ | $-5$ | $-2$ | $1$ | $4$ | $7$ |

**Solution**: The model is related to a linear function because the first difference is the same value:

$y \rightarrow -2 - (-5) = 3, 1 - (-2) = 3, 4 - 1 = 3, 7 - 4 = 3.$

$x \rightarrow -3 - (-5) = 2, -1 - (-3) = 2, 1 - (-1) = 2, 3 - 1 = 2.$

# Linear vs Exponential Growth

- The difference between linear and exponential growth is related to the difference in the $y$ values change when the $x$ values increase by a constant amount. In such a way that:
    - A relationship is a linear growth if the $y$ values have equal differences.
    - A relationship is an exponential growth if the $y$ values have an equal ratio.
    - Otherwise, it is neither.
- Actually, a linear growth diagram is similar to a straight line and an exponential growth diagram is similar to an exponential function.

- Much faster than linear growth is exponential growth.

## Examples:

**Example 1.** If 75 students graduate from high school every year. Is the relationship linear, exponential, or neither?
List the data as follows: (1,75), (2,150), (3,225), (4,300) and ⋯.

*Solution:* Considering that in each year first components increase by exactly 1 unit and the value of the second component increases with a constant difference of 75. It is Linear growth.

**Example 2.** Using the data in this table, determine whether this relationship is linear, exponential, or neither.

| $x$ | 0 | 1 | 2 | 3 | 4 | 5 |
|-----|---|---|----|----|----|----|
| $y$ | 3 | 6 | 12 | 24 | 48 | 96 |

*Solution:* We know that the $x$ value increase by 1 unit and the $y$ value increase by the constant ratio of 2. Then, this relationship is exponential.

**Example 3.** According to the following diagram, determine whether this relationship is linear, exponential, or neither.

*Solution:* Here, the $x$ value increase by 3 unit, however, the difference between the $y$ values are not constant and the ratios aren't constant. Therefore, it is neither.

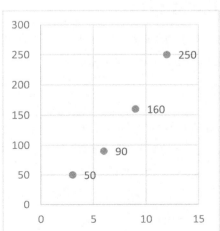

# Chapter 8: Practices

✎ **Sketch the graph of each function.**

1) $y = 2 \cdot \left(\frac{1}{2}\right)^x$

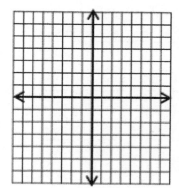

2) $y = 4 \cdot (2)^x$

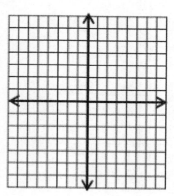

✎ **Determine whether the following table of values represents a linear function, an exponential function, or a quadratic function.**

3) _____

| $x$ | $-2$ | $-1$ | 0 | 1 | 2 |
|---|---|---|---|---|---|
| $y$ | $\frac{1}{2}$ | 1 | 2 | 4 | 8 |

4) _____

| $x$ | $-2$ | $-1$ | 0 | 1 | 2 |
|---|---|---|---|---|---|
| $y$ | 5 | 2 | 1 | 2 | 5 |

5) _____

| $x$ | $-2$ | $-1$ | 0 | 1 | 2 |
|---|---|---|---|---|---|
| $y$ | $-4$ | $-1$ | 2 | 5 | 8 |

✎ **Using the data in this table, determine whether this relationship is linear, exponential, or neither.**

6) _____

| $x$ | $-2$ | $-1$ | 0 | 1 | 2 |
|---|---|---|---|---|---|
| $y$ | $-6$ | $-3$ | 0 | 3 | 6 |

**Effortless**
**Math**
**Education**

# Chapter 8: Answers

1)

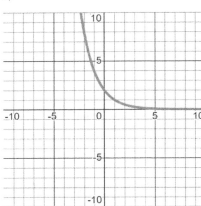

2)

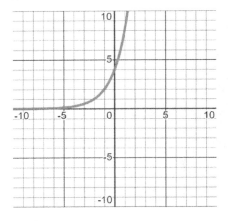

3) Exponential

4) Quadratic

5) Linear

6) $y = 3x$, Linear growth

# CHAPTER

# 9 Logarithms

Math topics that you'll learn in this chapter:

- ☑ Evaluating Logarithms
- ☑ Properties of Logarithms
- ☑ Natural Logarithms
- ☑ Solving Logarithmic Equations

121

# Evaluating Logarithms

- Logarithm is another way of writing exponent. $log_b y = x$ is equivalent to $y = b^x$.

- Learn some logarithms rules: ($a > 0$, $a \neq 0$, $M > 0$, $N > 0$, and $k$ is a real number.)

Rule 1: $log_a(M \cdot N) = log_a M + log_a N$,    Rule 4: $log_a a = 1$,

Rule 2: $log_a \frac{M}{N} = log_a M - log_a N$,    Rule 5: $log_a 1 = 0$,

Rule 3: $log_a M^k = k \, log_a M$,    Rule 6: $a^{log_a k} = k$.

## Examples:

**Example 1.** Evaluate: $log_2 32$.

**Solution**: Rewrite 32 in power base form: $32 = 2^5$, then:

$log_2 32 = log_2(2)^5$.

Use $log$ rule: $log_a M^k = k \, log_a M \rightarrow log_2(2)^5 = 5 \, log_2 2$.

Use $log$ rule: $log_a a = 1 \rightarrow log_2 2 = 1 \rightarrow 5 \, log_2 2 = 5$.

Therefore: $log_2 32 = 5$.

**Example 2.** Evaluate: $3 \, log_5 125$.

**Solution**: Rewrite 125 in power base form: $125 = 5^3$, then:

$log_5 125 = log_5(5)^3$.

Use $log$ rule: $log_a M^k = k \, log_a M \rightarrow log_5(5)^3 = 3 \, log_5 5$.

Use $log$ rule: $log_a a = 1 \rightarrow log_5 5 = 1 \rightarrow 3 \, log_5 5 = 3$.

Therefore: $3 \, log_5 125 = 3 \times 3 = 9$.

**Example 3.** Evaluate: $log_3(3)^5$

**Solution**: Use $log$ rule: $log_a M^k = k \, log_a M \rightarrow log_3(3)^5 = 5 \, log_3 3$.

Use $log$ rule: $log_a a = 1 \rightarrow log_3 3 = 1 \rightarrow 5 \, log_3 3 = 5 \times 1 = 5$.

# Properties of Logarithms

- Using some of the properties of logs, (the product rule, quotient rule, and power rule) sometimes we can expand a logarithm expression (expanding) or convert some logarithm expressions into a single logarithm (condensing).

- Let's review some logarithms properties:

$$a^{log_a b} = b$$

$$log_a 1 = 0$$

$$log_a a = 1$$

$$log_a(x.y) = log_a x + log_a y$$

$$log_a \frac{x}{y} = log_a x - log_a y$$

$$log_a \frac{1}{x} = - log_a x$$

$$log_a x^p = p \, log_a x$$

$$log_{a^k} x = \frac{1}{k} log_a x, \text{ for } k \neq 0$$

$$log_a x = log_{a^c} x^c$$

$$log_a x = \frac{1}{log_x a}$$

## Examples:

**Example 1.** Expand this logarithm. $log_a(3 \times 5) =$

**Solution**: Use $log$ rule: $log_a(x \cdot y) = log_a x + log_a y$.
Then:
$log_a(3 \times 5) = log_a 3 + log_a 5$.

**Example 2.** Condense this expression to a single logarithm. $log_a 2 - log_a 7$

**Solution**: Use $log$ rule: $log_a x - log_a y = log_a \frac{x}{y}$.

Then:
$log_a 2 - log_a 7 = log_a \frac{2}{7}$.

**Example 3.** Expand this logarithm. $log \left(\frac{1}{7}\right) =$

**Solution**: Use $log$ rule: $log_a \frac{1}{x} = - log_a x$.

Then:
$log \left(\frac{1}{7}\right) = - log \, 7$.

# Natural Logarithms

- A natural logarithm is a logarithm that has a special base of the mathematical constant $e$, which is an irrational number approximately equal to 2.71.

- The natural logarithm of $x$ is generally written as $ln\, x$, or $log_e\, x$.

## Examples:

**Example 1.** Expand this natural logarithm. $ln\, 4x^2 =$

*Solution*: Use *log* rule: $log_a(x.y) = log_a\, x + log_a\, y$.
Therefore:
$ln\, 4x^2 = ln\, 4 + ln\, x^2$.
Now, use *log* rule: $log_a(M)^k = k. log_a(M)$.
Then:
$ln\, 4 + ln\, x^2 = ln\, 4 + 2\, ln\, x$.

**Example 2.** Condense this expression to a single logarithm. $ln\, x - log_e\, 2y$

*Solution*: Use *log* rule: $log_a\, x - log_a\, y = log_a\frac{x}{y}$.

Then:
$ln\, x - log_e\, 2y = ln\frac{x}{2y}$.

**Example 3.** Solve this equation for $x$: $e^x = 6$.

*Solution*: If $f(x) = g(x)$, then: $ln\big(f(x)\big) = ln\big(g(x)\big) \rightarrow ln(e^x) = ln(6)$.
Use *log* rule: $log_a\, x^b = b\, log_a\, x \rightarrow ln(e^x) = x\, ln(e) \rightarrow xln(e) = ln(6)$.
We know that: $ln(e) = 1$, then:
$x = ln(6)$.

**Example 4.** Solve this equation for $x$: $ln(4x - 2) = 1$.

*Solution*: Use *log* rule: $a = log_b(b^a) \rightarrow 1 = ln(e^1) = ln(e) \rightarrow ln(4x - 2) = ln(e)$.
When the *log*s have the same base: $log_b\big(f(x)\big) = log_b\big(g(x)\big) \rightarrow f(x) = g(x)$.
$ln(4x - 2) = ln(e)$, then:
$4x - 2 = e \rightarrow x = \frac{e+2}{4}$.

# Solving Logarithmic Equations

**To solve a logarithm equation:**

- Convert the logarithmic equation to an exponential equation when it's possible. (If no base is indicated, the base of the logarithm is 10.)

- Condense logarithms if you have more than one *log* on one side of the equation.

- Plug the answers back into the original equation and check to see if the solution works.

## Examples:

**Example 1.** Find the value of $x$ in this equation. $log_2(36 - x^2) = 4$

*Solution*: Use *log* rule: $log_b x = log_b y$, then: $x = y$.

We can write number 4 as a logarithm: $4 = log_2(2^4)$.
Then:
$log_2(36 - x^2) = log_2(2^4) = log_2 16$.
Then: $36 - x^2 = 16 \rightarrow 36 - 16 = x^2 \rightarrow x^2 = 20 \rightarrow x = \pm\sqrt{20} = \pm 2\sqrt{5}$.
You can plug in back the solutions into the original equation to check your answer.
$x = \sqrt{20} \rightarrow log_2\left(36 - \left(\sqrt{20}\right)^2\right) = 4 \rightarrow log_2(36 - 20) = 4 \rightarrow log_2 16 = 4$.
$x = -\sqrt{20} \rightarrow log_2\left(36 - \left(-\sqrt{20}\right)^2\right) = 4 \rightarrow log_2(36 - 20) = 4 \rightarrow log_2 16 = 4$.
Both solutions work in the original equation.

**Example 2.** Find the value of $x$ in this equation. $log(5x + 2) = log(3x - 1)$

*Solution*: When the *log*s have the same base: $f(x) = g(x)$, then:

$ln\big(f(x)\big) = ln\big(g(x)\big)$, $log(5x + 2) = log(3x - 1) \rightarrow 5x + 2 = 3x - 1$. Then:
$5x + 2 - 3x + 1 = 0 \rightarrow 2x + 3 = 0 \rightarrow 2x = -3 \rightarrow x = -\frac{3}{2}$.
Verify Solution: $log\left(5\left(-\frac{3}{2}\right) + 2\right) = log(-5.5)$.

Logarithms of negative numbers are not defined. Therefore, there is no solution for this equation.

# Chapter 9: Practices

### ✍ Expand each logarithm.

1) $log_b(2 \times 9) =$

2) $log_b(5 \times 7) =$

3) $log_b(xy) =$

4) $log_b(2x^2 \times 3y) =$

### ✍ Evaluate each logarithm.

5) $2\,log_9(9) =$

6) $3\,log_2(8) =$

7) $2\,log_5(125) =$

8) $log_{100}(1) =$

9) $log_{10}(100) =$

10) $3\,log_4(16) =$

11) $\frac{1}{2}\,log_3(81) =$

12) $log_7(343) =$

### ✍ Reduce the following expressions to simplest form.

13) $e^{\ln 4 + \ln 5} =$

14) $e^{\ln\left(\frac{9}{e}\right)} =$

15) $e^{\ln 2 + \ln 7} =$

16) $6\,ln(e^5) =$

### ✍ Find the value of the variables in each equation.

17) $log_3 8x = 3 \rightarrow x = $ ____

18) $log_4 2x = 5 \rightarrow x = $ ____

19) $log_4 5x = 0 \rightarrow x = $ ____

20) $log\, 4x = log\, 5 \rightarrow x = $ ____

**Effortless
Math
Education**

# Chapter 9: Answers

1) $log_b 2 + 2 log_b 3$

2) $log_b 5 + log_b 7$

3) $log_b x + log_b y$

4) $log_b 6 + 2 log_b x + log_b y$

5) 2

6) 9

7) 6

8) 0

9) 2

10) 6

11) 2

12) 3

13) 20

14) $\frac{9}{e}$

15) 14

16) 30

17) $\frac{27}{8}$

18) 512

19) $\frac{1}{5}$

20) $\frac{5}{4}$

# CHAPTER

# 10 Radical Expressions

Math topics that you'll learn in this chapter:

- ☑ Simplifying Radical Expressions
- ☑ Simplifying Radical Expressions Involving Fractions
- ☑ Multiplying Radical Expressions
- ☑ Adding and Subtracting Radical Expressions
- ☑ Domain and Range of Radical Functions
- ☑ Radical Equations
- ☑ Solving Radical Inequalities

129

# Simplifying Radical Expressions

- Find the prime factors of the numbers or expressions inside the radical.

- Use radical properties to simplify the radical expression:

$$\sqrt[n]{x^a} = x^{\frac{a}{n}}, \quad \sqrt[n]{xy} = x^{\frac{1}{n}} \times y^{\frac{1}{n}}, \quad \sqrt[n]{\frac{x}{y}} = \frac{x^{\frac{1}{n}}}{y^{\frac{1}{n}}}, \text{ and } \sqrt[n]{x} \times \sqrt[n]{y} = \sqrt[n]{xy}.$$

## Examples:

**Example 1.** Find the square root of $\sqrt{144x^2}$.

**Solution**: Find the factor of the expression $144x^2$: $144 = 12 \times 12$ and $x^2 = x \times x$.
Now use the radical rule: $\sqrt[n]{a^n} = a$.
Then: $\sqrt{12^2} = 12$ and $\sqrt{x^2} = x$.
Finally:
$\sqrt{144x^2} = \sqrt{12^2} \times \sqrt{x^2} = 12 \times x = 12x$.

**Example 2.** Write this radical in exponential form. $\sqrt[3]{x^4}$

**Solution**: To write a radical in exponential form, use this rule: $\sqrt[n]{x^a} = x^{\frac{a}{n}}$.
Then:
$\sqrt[3]{x^4} = x^{\frac{4}{3}}$.

**Example 3.** Simplify. $\sqrt{8x^3}$

**Solution**: First factor the expression $8x^3$: $8x^3 = 2^3 \times x \times x \times x$.
We need to find perfect squares: $8x^3 = 2^2 \times 2 \times x^2 \times x = 2^2 \times x^2 \times 2x$.
Then: $\sqrt{8x^3} = \sqrt{2^2 \times x^2} \times \sqrt{2x}$.
Now use the radical rule: $\sqrt[n]{a^n} = a$. Then:
$\sqrt{2^2 \times x^2} \times \sqrt{(2x)} = 2x \times \sqrt{2x} = 2x\sqrt{2x}$.

**Example 4.** Simplify. $\sqrt{27a^5b^4}$

**Solution**: First factor the expression $27a^5b^4$: $27a^5b^4 = 3^3 \times a^5 \times b^4$.
We need to find perfect squares: $27a^5b^4 = 3^2 \times 3 \times a^4 \times a \times b^4$. Then:
$\sqrt{27a^5b^4} = \sqrt{3^2 \times a^4 \times b^4} \times \sqrt{3a}$.

Now use the radical rule: $\sqrt[n]{a^n} = a$. Then:
$\sqrt{3^2 \times a^4 \times b^4} \times \sqrt{3a} = 3 \times a^2 \times b^2 \times \sqrt{3a} = 3a^2b^2\sqrt{3a}$.

# Simplifying Radical Expressions Involving Fractions

- Radical expressions cannot be in the denominator. (Number in the bottom)

- To get rid of the radical in the denominator, multiply both the numerator and denominator by the radical in the denominator.

- If there is a radical and another integer in the denominator, multiply both the numerator and denominator by the conjugate of the denominator.

- The conjugate of $a + b$ is $a - b$ and vice versa.

## Examples:

**Example 1.** Simplify. $\frac{2}{\sqrt{3}-2}$

**Solution**: Multiply by the conjugate:

$\frac{\sqrt{3}+2}{\sqrt{3}+2} \rightarrow \frac{2}{\sqrt{3}-2} \times \frac{\sqrt{3}+2}{\sqrt{3}+2}.$

Since $(\sqrt{3} - 2)(\sqrt{3} + 2) = -1$,

Then:

$\frac{2}{\sqrt{3}-2} = \frac{2(\sqrt{3}+2)}{-1}.$

Use the fraction rule:

$\frac{a}{-b} = -\frac{a}{b} \rightarrow \frac{2(\sqrt{3}+2)}{-1} = -\frac{2(\sqrt{3}+2)}{1} = -2(\sqrt{3} + 2).$

**Example 2.** Simplify. $\frac{3}{\sqrt{7}-2}$

**Solution**: Multiply by the conjugate: $\frac{\sqrt{7}+2}{\sqrt{7}+2}.$

$\frac{\sqrt{7}+2}{\sqrt{7}+2} \rightarrow \frac{3}{\sqrt{7}-2} \times \frac{\sqrt{7}+2}{\sqrt{7}+2}.$

Therefore:

$\frac{3}{\sqrt{7}-2} \times \frac{\sqrt{7}+2}{\sqrt{7}+2} = \frac{3(\sqrt{7}+2)}{3} \rightarrow \frac{3(\sqrt{7}+2)}{3} = \sqrt{7} + 2.$

Finally:

$\frac{3}{\sqrt{7}-2} = \sqrt{7} + 2.$

# Multiplying Radical Expressions

To multiply radical expressions:

- Multiply the numbers outside of the radicals.

- Multiply the numbers inside the radicals.

- Simplify if needed.

## Examples:

**Example 1.** Evaluate. $\sqrt{16} \times \sqrt{9} =$

*Solution*: First factor the numbers: $16 = 4^2$ and $9 = 3^2$.

Then:

$\sqrt{16} \times \sqrt{9} = \sqrt{4^2} \times \sqrt{3^2}$.

Now use the radical rule: $\sqrt[n]{a^n} = a$.

Then:

$\sqrt{4^2} \times \sqrt{3^2} = 4 \times 3 = 12$.

**Example 2.** Evaluate. $2\sqrt{5} \times 3\sqrt{2} =$

*Solution*: Multiply the numbers: $2 \times 3 = 6$.

$2\sqrt{5} \times 3\sqrt{2} = 6\sqrt{5}\sqrt{2}$.

Use the radical rule:

$\sqrt{a}\sqrt{b} = \sqrt{ab} \rightarrow 6\sqrt{5}\sqrt{2} = 6\sqrt{5 \times 2} = 6\sqrt{10}$.

**Example 3.** Evaluate. $5\sqrt{12a^3b^3} \cdot \sqrt{3ab^2}$

*Solution*: First multiply the expression inside the radicals:

$12a^3b^3 \times 3ab^2 = 36a^4b^5$.

Now factor the expression:

$36a^4b^5 = 3^2 \times 2^2 \times a^4 \times b^5$.

We need to find the perfect square: $36a^4b^5 = 3^2 \times 2^2 \times a^4 \times b^4 \times b$.

Therefore: $\sqrt{12a^3b^3} \cdot \sqrt{3ab^2} = \sqrt{36a^4b^5} = \sqrt{3^2 \times 2^2 \times a^4 \times b^4 \times b} = 6a^2b^2\sqrt{b}$.

Finally:

$5\sqrt{12a^3b^3} \cdot \sqrt{3ab^2} = 30a^2b^2\sqrt{b}$.

# Adding and Subtracting Radical Expressions

- Only numbers that have the same radical part can be added or subtracted.

- Remember, combining "unlike" radical terms is not possible.

- For numbers with the same radical part, just add or subtract factors outside the radicals.

## Examples:

**Example 1.** Simplify. $4\sqrt{5} + 3\sqrt{5} =$

*Solution*: Add like terms:

$4\sqrt{5} + 3\sqrt{5} = 7\sqrt{5}$.

**Example 2.** Simplify. $2\sqrt{7} + 4\sqrt{7} =$

*Solution*: Add like terms:

$2\sqrt{7} + 4\sqrt{7} = 6\sqrt{7}$.

**Example 3.** Simplify. $5\sqrt{3} + 2\sqrt{3} =$

*Solution*: Add like terms:

$5\sqrt{3} + 2\sqrt{3} = 7\sqrt{3}$.

**Example 4.** Simplify. $3\sqrt{2} + 2\sqrt{5} + 5\sqrt{2} =$

*Solution*: Add like terms:

$3\sqrt{2} + 5\sqrt{2} = 8\sqrt{2}$.

Therefore:

$3\sqrt{2} + 2\sqrt{5} + 5\sqrt{2} = 8\sqrt{2} + 2\sqrt{5}$.

**Example 5.** Simplify. $2\sqrt{3} + \sqrt{3} + 5 =$

*Solution*: Add like terms:

$2\sqrt{3} + \sqrt{3} = 3\sqrt{3}$.

Therefore:

$2\sqrt{3} + \sqrt{3} + 5 = 3\sqrt{3} + 5$.

# Domain and Range of Radical Functions

- To find the domain and range of radical functions, remember that having a negative number under the square root symbol is not possible. (For square roots)

- To find the domain of the function, find all possible values of the variable inside radical.

- To find the range, plugin the minimum and maximum values of the variable inside radical.

## Examples:

**Example 1.** Find the domain and range of the radical function.

$$y = \sqrt{x - 2}$$

**Solution**: For the domain, find non-negative values for radicals: $x - 2 \geq 0$. The domain of functions: $x - 2 \geq 0 \rightarrow x \geq 2$.

Domain of the function $y = \sqrt{x - 2}$:

$x \geq 2$.

For the range, the range of a radical function of the form $c\sqrt{ax + b} + k$ is: $f(x) \geq k$.

For the function $y = \sqrt{x - 2}$, the value of $k$ is 0. Then:

$f(x) \geq 0$.

Range of the function $y = \sqrt{x - 2}$: $f(x) \geq 0$.

**Example 2.** Find the domain and range of the radical function.

$$y = 3\sqrt{2x - 3} + 1$$

**Solution**: For the domain, find non-negative values for radicals: $2x - 3 \geq 0$. The Domain of functions: $2x - 3 \geq 0 \rightarrow 2x \geq 3 \rightarrow x \geq \frac{3}{2}$.

Domain of the function $y = 3\sqrt{2x - 3} + 1$:

$x \geq \frac{3}{2}$.

For the range, the range of a radical function of the form $c\sqrt{ax + b} + k$ is: $f(x) \geq k$.

For the function $y = 3\sqrt{2x - 3} + 1$, the value of $k$ is 1. Then:

$f(x) \geq 1$.

Range of the function $y = 3\sqrt{2x - 3} + 1$: $f(x) \geq 1$.

# Radical Equations

- Isolate the radical on one side of the equation.

- Square both sides of the equation to remove the radical.

- Solve the equation for the variable.

- Plugin the answer into the original equation to avoid extraneous values.

## Examples:

**Example 1.** Solve. $\sqrt{x} - 8 = 12$

***Solution***: Add 8 to both sides:

$\sqrt{x} - 8 + 8 = 12 + 8 \rightarrow \sqrt{x} = 20$.

Square both sides:

$\left(\sqrt{x}\right)^2 = 20^2 \rightarrow x = 400$.

**Example 2.** Solve. $\sqrt{x + 2} = 6$

***Solution***: Square both sides:

$\left(\sqrt{x + 2}\right)^2 = 6^2 \rightarrow x + 2 = 36$.

Subtract 2 to both sides:

$x + 2 - 2 = 36 - 2 \rightarrow x = 34$.

**Example 3.** Solve. $2\sqrt{x - 5} + 1 = 4$

***Solution***: Subtract 1 to both sides:

$2\sqrt{x - 5} + 1 - 1 = 4 - 1 \rightarrow 2\sqrt{x - 5} = 3$.

Divide 2 to both sides:

$\left(2\sqrt{x - 5}\right) \div 2 = 3 \div 2 \rightarrow \sqrt{x - 5} = \frac{3}{2}$.

Square both sides:

$\left(\sqrt{x - 5}\right)^2 = \left(\frac{3}{2}\right)^2 \rightarrow x - 5 = \frac{9}{4} \rightarrow x = \frac{29}{4}$.

bit.ly/3sl2qln

Find more at

# Solving Radical Inequalities

- Radical inequality is an inequality that contains rational expressions.

- To solve a radical inequality, follow these steps:

> Step1: Write radical expressions on one side of the inequality and other expressions on the other side.
> Step2: Raise both sides of the inequality to the power equal to the index of the radical.
> Step3: Simplify the obtained inequality and solve.

- You can solve radical inequalities by graphing.

## Examples:

**Example 1.** Solve: $2\sqrt[3]{1-x} - 3 \geq 0$.

*Solution*: First, rewrite the inequality as follows:

$$2\sqrt[3]{1-x} - 3 \geq 0 \rightarrow 2\sqrt[3]{1-x} \geq 3 \rightarrow \sqrt[3]{1-x} \geq \frac{3}{2}.$$

Raise both sides of the inequality to the power of 3:

$$\sqrt[3]{1-x} \geq \frac{3}{2} \rightarrow \left(\sqrt[3]{1-x}\right)^3 \geq \left(\frac{3}{2}\right)^3 \rightarrow 1 - x \geq \frac{27}{8}.$$

Simplify and solve: $1 - x \geq \frac{27}{8} \rightarrow 1 - \frac{27}{8} \geq x \rightarrow x \leq -\frac{19}{8}$.

The final answer to this problem is $(-\infty, -\frac{19}{8}]$.

**Example 2.** Solve: $1 - \sqrt{x+3} < 0$.

*Solution*: First, rewrite the inequality as follows:

$$1 - \sqrt{x+3} < 0 \rightarrow 1 < \sqrt{x+3} \rightarrow \sqrt{x+3} > 1.$$

Raise both sides of the inequality to the power of 2:

$$\sqrt{x+3} > 1 \rightarrow \left(\sqrt{x+3}\right)^2 > (1)^2 \rightarrow x + 3 > 1.$$

Simplify and solve: $x + 3 > 1 \rightarrow x > -2$. On the other hand, for a radical expression with an even index in the form $\sqrt[2k]{f(x)}$ where $k \in N$, is $f(x) \geq 0$. It means that:

$$x + 3 \geq 0 \rightarrow x \geq -3.$$

The final answer to this problem is $(-2, +\infty) \cap [-3, +\infty) = (-2, +\infty)$.

# Chapter 10: Practices

> 🔖 **Simplify.**

    1) $\sqrt{256y} =$                3) $\sqrt{144a^2b} =$

    2) $\sqrt{900} =$                4) $\sqrt{36 \times 9} =$

> 🔖 **Simplify.**

    5) $3\sqrt{5} + 2\sqrt{5} =$          7) $5\sqrt{2} + 10\sqrt{18} =$

    6) $6\sqrt{3} + 4\sqrt{27} =$        8) $7\sqrt{2} - 5\sqrt{8} =$

> 🔖 **Evaluate.**

    9) $\sqrt{5} \times \sqrt{3} =$          11) $3\sqrt{5} \times \sqrt{9} =$

    10) $\sqrt{6} \times \sqrt{8} =$        12) $2\sqrt{3} \times 3\sqrt{7} =$

> 🔖 **Simplify.**

    13) $\frac{1}{\sqrt{3}-6} =$          15) $\frac{\sqrt{3}}{1-\sqrt{6}} =$

    14) $\frac{5}{\sqrt{2}+7} =$         16) $\frac{2}{\sqrt{3}+5} =$

**Effortless Math Education**

✏ **Solve for $x$.**

17) $\sqrt{x} + 2 = 9$

18) $3 + \sqrt{x} = 12$

19) $\sqrt{x} + 5 = 30$

20) $\sqrt{x} - 9 = 27$

21) $10 = \sqrt{x + 1}$

22) $\sqrt{x + 4} = 3$

✏ **Identify the domain and range of each function.**

23) $y = \sqrt{x + 2} - 1$            25) $y = \sqrt{x - 4}$

24) $y = \sqrt{x + 1}$               26) $y = \sqrt{x - 3} + 1$

✏ **Solve.**

27) $3\sqrt{x} - 4 \geq 5$          30) $4\sqrt{x} + 6 \leq 8$

28) $1 - \sqrt{x + 8} < 0$       31) $\sqrt[3]{x - 3} \leq 6$

29) $\sqrt{x - 5} < 3$           32) $-3\sqrt[3]{x + 2} \leq 12$

**Effortless**
**Math**
**Education**

# Chapter 10: Answers

1) $16\sqrt{y}$

2) $30$

3) $12a\sqrt{b}$

4) $18$

5) $5\sqrt{5}$

6) $18\sqrt{3}$

7) $35\sqrt{2}$

8) $-3\sqrt{2}$

9) $\sqrt{15}$

10) $4\sqrt{3}$

11) $9\sqrt{5}$

12) $6\sqrt{21}$

13) $-\frac{\sqrt{3}+6}{33}$

14) $-\frac{5(\sqrt{2}-7)}{47}$

15) $-\frac{\sqrt{3}+3\sqrt{2}}{5}$

16) $-\frac{\sqrt{3}-5}{11}$

17) $x = 49$

18) $x = 81$

19) $x = 625$

20) $x = 1,296$

21) $x = 99$

22) $x = 5$

23) $x \geq -2, f(x) \geq -1$

24) $x \geq -1, f(x) \geq 0$

25) $x \geq 4, f(x) \geq 0$

26) $x \geq 3, f(x) \geq 1$

27) $x \geq 9$

28) $x > -7$

29) $5 \leq x < 14$

30) $0 \leq x \leq \frac{1}{4}$

31) $x \leq 219$

32) $x \geq -66$

**Effortless Math Education**

# 11 Rational and Irrational Expressions

Math topics that you'll learn in this chapter:

- ☑ Rational and Irrational Numbers
- ☑ Simplifying Rational Expressions
- ☑ Graphing Rational Expressions
- ☑ Multiplying Rational Expressions
- ☑ Dividing Rational Expressions
- ☑ Adding and Subtracting Rational Expressions
- ☑ Rational Equations
- ☑ Simplifying Complex Fractions
- ☑ Maximum and Minimum Points
- ☑ Solving Rational Inequalities
- ☑ Irrational Functions
- ☑ Direct, Inverse, Joint, and Combined Variation

141

# Rational and Irrational Numbers

- Rational numbers are actually fractions (not all fractions) that have positive and negative signs. In fact, integers, natural numbers, and arithmetic numbers are all subsets of rational numbers. Decimal numbers can be considered rational numbers because every decimal number can be written as a fraction whose denominator is one of the positive powers of 10 and whose form is an integer. Rational numbers are the result of dividing two numbers (Dividing an integer by a natural number). There are infinite fractions between two rational numbers. Rational numbers extend from negative infinity to positive infinity. Also, the symbol of rational numbers is $Q$.

- An irrational number within the system of numeration is defined as a real number that's not rational, that is, it can't be written as a fraction whose numerator and denominator are integers. An irrational number could be a decimal number whose decimal digits are endless but do not have any cycle period. The foremost famous of those numbers are $\pi$, $e$, and $\sqrt{2}$.

## Examples:

**Example 1.** Determine if $0.\overline{15}$ is a rational number or an irrational number.

*Solution*: All repeating decimals can be shown as the fraction of two integers and a rational number is a number that can be defined as the fraction of two integers. We know $0.\overline{15}$ can be expressed as $\frac{15}{99}$. So, $0.\overline{15}$ is a rational number.

**Example 2.** Determine if $\sqrt{7}$ is a rational number or an irrational number.

*Solution*: Irrational numbers can't be written as a fraction whose numerator and denominator are integers. Since $\sqrt{7}$ is equal to $2.645751\ldots$, it's a decimal number whose decimal digits are endless and it can't be written as a fraction whose numerator and denominator are integers. So, $\sqrt{7}$ is an irrational number.

**Example 3.** Determine if $-4.8$ is a rational number or an irrational number.

*Solution*: $-4.8$ is a rational number because it can be written as a fraction: $-\frac{48}{10}$.

bit.ly/3GzXFq2

Find more at

# Simplifying Rational Expressions

- Factorize the numerator and denominator if they are factorable.

- Find common factors of both the numerator and denominator.

- Remove the common factor in both the numerator and denominator.

- Simplify if needed.

## Examples:

**Example 1.** Simplify. $\frac{9x^2y}{3y^2}$

***Solution***: Cancel the common factor 3: $\frac{9x^2y}{3y^2} = \frac{3x^2y}{y^2}$.

Cancel the common factor $y$: $\frac{3x^2y}{y^2} = \frac{3x^2}{y}$.

Then:

$\frac{9x^2y}{3y^2} = \frac{3x^2}{y}$.

**Example 2.** Simplify. $\frac{x^2+5x-6}{x+6}$

***Solution***: Factor $x^2 + 5x - 6 = (x - 1)(x + 6)$.

Then: $\frac{x^2+5x-6}{x+6} = \frac{(x-1)(x+6)}{x+6}$.

Cancel the common factor: $(x + 6)$.

Then:

$\frac{(x-1)(x+6)}{x+6} = x - 1$.

**Example 3.** Simplify. $\frac{18a^3(b+1)c^2}{6a^2c(b^2-1)}$

***Solution***: Cancel common factor in numerator and denominator: $a^2$, $c$ and 6.

Then: $\frac{18a^3(b+1)c^2}{6a^2c(b^2-1)} = \frac{3a(b+1)c}{(b^2-1)}$.

Now, by factoring $b^2 - 1 = (b + 1)(b - 1)$, and cancel common factor $(b + 1)$, we have:

$\frac{3a(b+1)c}{(b^2-1)} = \frac{3a(b+1)c}{(b+1)(b-1)} = \frac{3ac}{b-1} \rightarrow \frac{18a^3(b+1)c^2}{6a^2c(b^2-1)} = \frac{3ac}{b-1}$.

# Graphing Rational Expressions

- A rational expression is a fraction in which the numerator and/or the denominator are polynomials. Examples: $\frac{1}{x}$, $\frac{x^2}{x-1}$, $\frac{x^2-x+2}{x^2+5x+1}$, $\frac{m^2+6m-5}{m-2m}$.

- To graph a rational function:

   o Find the vertical asymptotes of the function if there are any. (Vertical asymptotes are vertical lines that correspond to the zeroes of the denominator. The graph will have a vertical asymptote at $x = a$ if the denominator is zero at $x = a$ and the numerator isn't zero at $x = a$.)

   o Find the horizontal or slant asymptote. (If the numerator has a bigger degree than the denominator, there will be a slant asymptote. To find the slant asymptote, divide the numerator by the denominator using either long division or synthetic division.)

   o If the denominator has a bigger degree than the numerator, the horizontal asymptote is the $x-$axis or the line $y = 0$. If they have the same degree, the horizontal asymptote equals the leading coefficient (the coefficient of the largest exponent) of the numerator divided by the leading coefficient of the denominator.

   o Find intercepts and plug in some values of $x$ and solve for $y$, then graph the function.

## Example:

Graph rational function. $f(x) = \frac{x^2-x+2}{x-1}$.

*Solution*: First, notice that the graph is in two pieces. Most rational functions have graphs in multiple pieces. Find the $y-$intercept by substituting zero for $x$ and solving for $y$. $f(x)$:

$x = 0 \rightarrow y = \frac{x^2-x+2}{x-1} = \frac{0^2-0+2}{0-1} = -2$,

$y-$intercept: $(0, -2)$.

Asymptotes of $\frac{x^2-x+2}{x-1}$: Vertical: $x = 1$, Slant asymptote: $y = x$ (Divide the numerator by the denominator). After finding the asymptotes, you can plug in some values for $x$ and solve for $y$. Here is the sketch for this function.

# Multiplying Rational Expressions

- Multiplying rational expressions is the same as multiplying fractions. First, multiply numerators and then multiply denominators. Then, simplify as needed.

## Examples:

**Example 1.** Solve: $\frac{x+6}{x-1} \times \frac{x-1}{5} =$.

*Solution*: Multiply numerators and denominators:

$\frac{a}{b} \times \frac{c}{d} = \frac{a \times c}{b \times d}$.

Then:

$\frac{x+6}{x-1} \times \frac{x-1}{5} = \frac{(x+6)(x-1)}{5(x-1)}$.

Cancel the common factor: $(x - 1)$.

Therefore:

$\frac{(x+6)(x-1)}{5(x-1)} = \frac{(x+6)}{5}$.

**Example 2.** Solve: $\frac{x-2}{x+3} \times \frac{2x+6}{x-2} =$.

*Solution*: Multiply numerators and denominators:

$\frac{a}{b} \times \frac{c}{d} = \frac{a \times c}{b \times d}$.

So, we have:

$\frac{x-2}{x+3} \times \frac{2x+6}{x-2} = \frac{(x-2)(2x+6)}{(x+3)(x-2)}$.

Cancel the common factor:

$\frac{(x-2)(2x+6)}{(x+3)(x-2)} = \frac{(2x+6)}{(x+3)}$.

Factor $2x + 6 = 2(x + 3)$.

Then:

$\frac{2(x+3)}{(x+3)} = 2$.

# Dividing Rational Expressions

- To divide rational expressions, use the same method we use for dividing fractions. (Keep, Change, Flip)

- Keep the first rational expression, change the division sign to multiplication, and flip the numerator and denominator of the second rational expression. Then, multiply numerators and multiply denominators. Simplify as needed.

## Examples:

**Example 1.** Solve. $\frac{x+2}{3x} \div \frac{x^2+5x+6}{3x^2+3x} =$

**Solution**: Use fractions division rule: $\frac{a}{b} \div \frac{c}{d} = \frac{a}{b} \times \frac{d}{c} = \frac{a \times d}{b \times c}$.

Therefore:

$\frac{x+2}{3x} \div \frac{x^2+5x+6}{3x^2+3x} = \frac{x+2}{3x} \times \frac{3x^2+3x}{x^2+5x+6} = \frac{(x+2)(3x^2+3x)}{(3x)(x^2+5x+6)}$.

Now, factorize the expressions $3x^2 + 3x$ and $(x^2 + 5x + 6)$. Then:

$3x^2 + 3x = 3x(x + 1)$ and $x^2 + 5x + 6 = (x + 2)(x + 3)$.

Simplify: $\frac{(x+2)(3x^2+3x)}{(3x)(x^2+5x+6)} = \frac{(x+2)(3x)(x+1)}{(3x)(x+2)(x+3)}$, cancel common factors.

Then:

$\frac{(x+2)(3x)(x+1)}{(3x)(x+2)(x+3)} = \frac{x+1}{x+3}$.

**Example 2.** Solve. $\frac{5x}{x+3} \div \frac{x}{2x+6} =$

**Solution**: Use fractions division rule: $\frac{a}{b} \div \frac{c}{d} = \frac{a}{b} \times \frac{d}{c} = \frac{a \times d}{b \times c}$.

Then:

$\frac{5x}{x+3} \div \frac{x}{2x+6} = \frac{5x}{x+3} \times \frac{2x+6}{x} = \frac{5x(2x+6)}{x(x+3)} = \frac{5x \times 2(x+3)}{x(x+3)}$.

Cancel common factor:

$\frac{5x \times 2(x+3)}{x(x+3)} = \frac{10x(x+3)}{x(x+3)} = 10$.

# Adding and Subtracting Rational Expressions

For adding and subtracting rational expressions:

- Find the least common denominator (LCD).

- Write each expression using the LCD.

- Add or subtract the numerators.

- Simplify as needed.

## Examples:

**Example 1.** Solve. $\frac{4}{2x+3} + \frac{x-2}{2x+3} =$

*Solution*: The denominators are equal. Then, use the fractions addition rule:
$\frac{a}{c} \pm \frac{b}{c} = \frac{a \pm b}{c}$.
Therefore:
$\frac{4}{2x+3} + \frac{x-2}{2x+3} = \frac{4+(x-2)}{2x+3} = \frac{x+2}{2x+3}$.

**Example 2.** Solve. $\frac{x+4}{x-5} + \frac{x-4}{x+6} =$

*Solution*: Find the least common denominator of $x-5$ and $x+6$: $(x-5)(x+6)$.

$\frac{x+4}{x-5} + \frac{x-4}{x+6} = \frac{(x+4)(x+6)}{(x-5)(x+6)} + \frac{(x-4)(x-5)}{(x+6)(x-5)} = \frac{(x+4)(x+6)+(x-4)(x-5)}{(x+6)(x-5)}$.

Expand:

$(x+4)(x+6) + (x-4)(x-5) = 2x^2 + x + 44$.

Then: $\frac{(x+4)(x+6)+(x-4)(x-5)}{(x+6)(x-5)} = \frac{2x^2+x+44}{(x+6)(x-5)} = \frac{2x^2+x+44}{x^2+x-30}$.

**Example 3.** Simplify the expression: $\frac{1}{x-3} + \frac{1}{x+1}$.

*Solution*: Find the least common denominator of $x-3$ and $x+1$: $(x-3)(x+1)$.

$\frac{1}{x-3} + \frac{1}{x+1} = \frac{(x+1)}{(x-3)(x+1)} + \frac{(x-3)}{(x+1)(x-3)} = \frac{(x+1)+(x-3)}{(x+1)(x-3)} = \frac{2x-2}{x^2-2x-3}$.

Therefore: $\frac{1}{x-3} + \frac{1}{x+1} = \frac{2x-2}{x^2-2x-3}$.

bit.ly/3d305i2

Find more at

# Rational Equations

For solving rational equations, we can use the following methods:

- **Converting to a common denominator:** In this method, you need to get a common denominator for both sides of the equation. Then, make numerators equal and solve for the variable.

- **Cross-multiplying:** This method is useful when there is only one fraction on each side of the equation. Simply multiply the first numerator by the second denominator and make the result equal to the product of the second numerator and the first denominator.

## Examples:

**Example 1.** Solve. $\frac{x-2}{x+1} = \frac{x+4}{x-2}$

**Solution:** Use cross multiply method: If $\frac{a}{b} = \frac{c}{d}$, then: $a \times d = b \times c$.

$\frac{x-2}{x+1} = \frac{x+4}{x-2} \rightarrow (x-2)(x-2) = (x+4)(x+1)$.

Expand: $(x-2)^2 = x^2 - 4x + 4$ and $(x+4)(x+1) = x^2 + 5x + 4$.

Then:

$x^2 - 4x + 4 = x^2 + 5x + 4$.

Now, simplify:

$x^2 - 4x = x^2 + 5x$.

Subtract both sides $(x^2 + 5x)$. Then:

$x^2 - 4x - (x^2 + 5x) = x^2 + 5x - (x^2 + 5x) \rightarrow -9x = 0 \rightarrow x = 0$.

**Example 2.** Solve. $\frac{2x}{x-3} = \frac{2x+2}{2x-6}$

**Solution:** Multiply the numerator and denominator of the rational expression on the left by 2 to get a common denominator $(2x - 6)$:

$\frac{2(2x)}{2(x-3)} = \frac{4x}{2x-6}$.

Now, the denominators on both side of the equation are equal.

Therefore, their numerators must be equal too.

$\frac{4x}{2x-6} = \frac{2x+2}{2x-6} \rightarrow 4x = 2x + 2 \rightarrow 2x = 2 \rightarrow x = 1$.

# Simplifying Complex Fractions

- Convert mixed numbers to improper fractions.

- Simplify all fractions.

- Write the fraction in the numerator of the main fraction line then write the division sign ($\div$) and the fraction of the denominator.

- Use the normal method for dividing fractions.

- Simplify as needed.

## Examples:

**Example 1.** Simplify. $\dfrac{\frac{3}{5}}{\frac{2}{25}-\frac{5}{16}}$

***Solution***: First, simplify the denominator: $\frac{2}{25}-\frac{5}{16}=-\frac{93}{400}$. Then:

$$\frac{\frac{3}{5}}{\frac{2}{25}-\frac{5}{16}}=\frac{\frac{3}{5}}{-\frac{93}{400}};$$

Now, write the complex fraction using the division sign:

$$\frac{\frac{3}{5}}{-\frac{93}{400}}=\frac{3}{5}\div\left(-\frac{93}{400}\right).$$

Use the dividing fractions rule: Keep, Change, Flip (Keep the first fraction, Change the division sign to multiplication, Flip the second fraction).

$$\frac{3}{5}\div\left(-\frac{93}{400}\right)=\frac{3}{5}\times-\frac{400}{93}=-\frac{240}{93}=-\frac{80}{31}=-2\frac{18}{31}.$$

**Example 2.** Simplify. $\dfrac{\frac{2}{5}\div\frac{1}{3}}{\frac{5}{9}+\frac{1}{3}}$

***Solution***: First, simplify the numerator: $\frac{2}{5}\div\frac{1}{3}=\frac{6}{5}$, then, simplify the denominator:

$$\frac{5}{9}+\frac{1}{3}=\frac{8}{9}.$$

Now, write the complex fraction using the division sign ($\div$):

$$\frac{\frac{2}{5}\div\frac{1}{3}}{\frac{5}{9}+\frac{1}{3}}=\frac{\frac{6}{5}}{\frac{8}{9}}=\frac{6}{5}\div\frac{8}{9}.$$

Use the dividing fractions rule:

(Keep, Change, Flip) $\frac{6}{5}\div\frac{8}{9}=\frac{6}{5}\times\frac{9}{8}=\frac{54}{40}=\frac{27}{20}=1\frac{7}{20}.$

bit.ly/3iu0Yql

Find more at

# Maximum and Minimum Points

- There are two types of maximum (minimum) points: absolute maximum (minimum) and local maximum (minimum).
- $f(a)$ is a local maximum if there is an interval around $a$ such that $f(a) > f(x)$ for all values of $x$ in the interval, where $x \neq a$.
- $f(a)$ is a local minimum if there is an interval around $a$ such that $f(a) < f(x)$ for all values of $x$ in the interval, where $x \neq a$.
- $f(a)$ is an absolute maximum if $f(a)$ is the largest value for all values of $x$ in the domain.
- $f(a)$ is an absolute minimum if $f(a)$ is the smallest value for all values of $x$ in the domain.

## Examples:

**Example 1.** Find the maximum points of the function: $f(x) = x^3 - 3x + 1$.

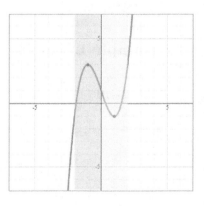

**Solution:** First draw the graph of the function. According to the graph of the function, the value 3 is a local maximum because in this value that corresponds to point $-1$ on the domain, there is an open interval like $(-2,0)$ around $-1$ such that $f(-1) > f(x)$ for all values of $x \in (-2,0)$. (Blue area). The value $-1$ is a local minimum of the function. Because in this value that corresponds to point 1 on the domain, there is an open interval like $(0,2)$ around $-1$ such that $f(1) < f(x)$ for all values of $x \in (0,2)$ (Pink area). The function does not have an absolute maximum or minimum, because the end of the function's behavior is infinite.

**Example 2.** Determine local and absolute maximum (minimum) points on the graph.

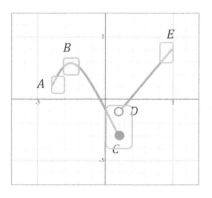

**Solution:** According to the graph, and the definition, it is enough to find an open interval around the points of the domain corresponding to the maximum (minimum) point, where the conditions of the maximum (minimum) point are satisfied on that interval.

Point $A$: The point is a local minimum. Because point $-4$ is on the domain and there is the open interval $(-4,-3)$ as such that the values of the function over this interval are all greater than the value of the function at point $-4$.

Point $B$: In the same way for point $B$, there is $(-3,-2)$ so that the value of the function on point $B$ is greater than the values of the function over this interval. Point $C$:

For point $C$, the value of the function on the graph at point $C$ is the lowest value on the domain of the function, therefore it is the absolute minimum point. Point $E$: Similarly, the point is an absolute maximum.

# Solving Rational Inequalities

- A rational inequality is an inequality that contains rational expressions.

- To solve a rational inequality, follow these steps:

  Step1: Write the inequality in a general form. It means to write the rational expression on the left side and zero on the right side of the inequality.

  Step2: Determine the points where the numerator and denominator expressions become zero. (Critical points)

  Step3: Using the points obtained in the previous step (Critical points), divide the set of all input values for the rational expression into smaller intervals.

  Step4: By putting an arbitrary point of each interval in the rational expression, determine Whether each interval is satisfied for the rational inequality.

- You can solve rational inequalities by graphing.

## Example:

Solve: $\frac{x^2+3}{x+3} < 2$.

**Solution**: First, write the inequality in general form:

$$\frac{x^2+3}{x+3} < 2 \rightarrow \frac{x^2+3}{x+3} - 2 < 0 \rightarrow \frac{x^2+3-2(x+3)}{x+3} < 0 \rightarrow \frac{x^2-2x-3}{x+3} < 0.$$

Determine the zeros of the numerator and denominator: $x^2 - 2x - 3 = 0$. So,

$(x + 1)(x - 3) = 0 \rightarrow x = -1$, and $x = 3$. In addition: $x + 3 = 0 \rightarrow x = -3$. Now, by putting an arbitrary point from each interval of the $(-\infty, -3)$, $(-3, -1)$, $(-1,3)$, and $(3, +\infty)$ in the inequality, we have:

$-4 \in (-\infty, -3) \rightarrow \frac{(-4)^2 - 2(-4) - 3}{(-4) + 3} < 0 \rightarrow -21 < 0$, this is true.

$-2 \in (-3, -1) \rightarrow \frac{(-2)^2 - 2(-2) - 3}{(-2) + 3} < 0 \rightarrow 5 < 0$, this is false.

$0 \in (-1,3) \rightarrow \frac{(0)^2 - 2(0) - 3}{(0) + 3} < 0 \rightarrow -1 < 0$, this is true.

$4 \in (3, +\infty) \rightarrow \frac{(4)^2 - 2(4) - 3}{(4) + 3} < 0 \rightarrow \frac{5}{7} < 0$, this is false.

The final answer to this problem is $(-\infty, -3) \cup (-1,3)$.

bit.ly/3CxBCPy

Find more at

# Irrational Functions

- An irrational function is a function that includes independent variables (expression) under the radical sign or a rational number for its exponent.

- An irrational function has expressions as follows:

$$f(x) = \sqrt[n]{(g(x))^m}, \text{ or } f(x) = (g(x))^{\frac{m}{n}}$$

Where $g(x)$ is a rational function?

  o If $n$ is odd, then $Domain(f) = Domain(g)$.

  o If $n$ is even, the domain of $f$ is equal to all values where $g(x) \geq 0$.

## Example:

What is the domain of the function: $y = \frac{\sqrt{x}}{1-\sqrt{x+2}}$?

*Solution:* Since the denominator expression must be non-zero, then:

$$1 - \sqrt{x+2} = 0 \rightarrow \sqrt{x+2} = 1 \rightarrow \left(\sqrt{x+2}\right)^2 = (1)^2 \rightarrow x + 2 = 1 \rightarrow x = -1.$$

This means that $-1$ is not in the domain of the function. Now check the radical expressions: So, for $\sqrt{x+2}$ because the radical index is even:

$$x + 2 \geq 0 \rightarrow x \geq -2.$$

Therefore, the denominator of the function accepts the values of the following interval: $[-2, +\infty) - \{-1\} = [-2, -1) \cup (-1, +\infty)$.

In addition, for $\sqrt{x}$: $x \geq 0$.

Here, the domain of the function $y = \frac{\sqrt{x}}{1-\sqrt{x+2}}$ is:

$$\{x \in [-2, +\infty): x \neq -1\} \cap \{x \in R: x \geq 0\} = [0, +\infty).$$

# Direct, Inverse, Joint, and Combined Variation

- There are different ways to express how one number changes relative to another number.

- Between the variables $x$, $y$, $z$ and a non-zero constant number $c$, have:

  o A direct variation relationship if $y = cx$. (As one number increases, so does the other)
  o An inverse variation relationship if $y = \frac{c}{x}$. (As one number increases, the other decreases)
  o A joint variation relationship if $y = cxz$.
  o A combined variation relationship if $y = c\left(\frac{x}{z}\right)$.

## Examples:

**Example 1.** When $x$ is 5 and $y$ is 3, find the constant of variation and an equation that inversely relates $y$ and $x$.

**Solution**: Considering the inverse variation relationship $y = \frac{c}{x} \rightarrow c = xy$, and substitute 5 for $x$ and 3 for $y$, we have: $c = (5) \cdot (3) \rightarrow c = 15$.

By putting the value of 15 for $c$ in the inverse variation relationship: $y = \frac{15}{x}$.

**Example 2.** Refer to the figure on the right. Write an equation to represent the area $S$, Identify the type of variation and the constant of variation.

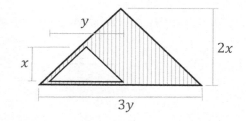

**Solution**: Use the triangle area to calculate the shaded region. We have:

$$S = \frac{1}{2}(2x) \cdot (3y) - \frac{1}{2}x \cdot y \rightarrow S = \frac{5}{2}xy$$

According to the definition of the joint variation relationship $y = cxz$, such that $c \neq 0$. Therefore, the area varies jointly as $x$ and $y$. The constant of variation is $\frac{5}{2}$.

# Chapter 11: Practices

## ✍ Determine if the number is rational or irrational number.

1) $\sqrt{25}$

2) $\sqrt{7}$

3) $4.8$

4) $24$

5) $90.790180\,.../$

6) $\frac{22}{38}$

7) $\sqrt{3}$

8) $2.514796\,.../$

## ✍ Simplify.

9) $\frac{16x^3}{20x^3} =$

10) $\frac{64x^3}{24x} =$

11) $\frac{25x^5}{15x^3} =$

12) $\frac{16}{2x-2} =$

13) $\frac{15x-3}{24} =$

14) $\frac{4x+16}{28} =$

15) $\frac{x^2-10x+25}{x-5} =$

16) $\frac{x^2-49}{x^2+3x-28} =$

17) $\frac{x^2+4x+4}{x^2-5x-14} =$

## ✍ Graph rational expressions.

18) $f(x) = \frac{x^2+2x-4}{x-2}$

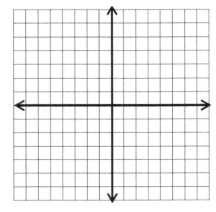

19) $f(x) = \frac{4x^3-16x+64}{x^2-2x-4}$

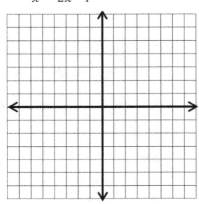

**Effortless
Math
Education**

## ✎ Simplify each expression.

20) $\dfrac{79x}{25} \cdot \dfrac{85}{27x^2} =$

21) $\dfrac{96}{38x} \cdot \dfrac{25}{45} =$

22) $\dfrac{84}{3} \cdot \dfrac{48x}{95} =$

23) $\dfrac{53}{43} \cdot \dfrac{46x^2}{31} =$

24) $\dfrac{93}{21x} \cdot \dfrac{34x}{51x} =$

25) $\dfrac{5x+50}{x+10} \cdot \dfrac{x-2}{5} =$

26) $\dfrac{x-7}{x+6} \cdot \dfrac{10x+60}{x-7} =$

27) $\dfrac{1}{x+10} \cdot \dfrac{10x+30}{x+3} =$

## ✎ Divide.

28) $\dfrac{12x}{3} \div \dfrac{5}{8} =$

29) $\dfrac{10x^2}{7} \div \dfrac{3x}{12} =$

30) $\dfrac{x+5}{5x^2-10x} \div \dfrac{1}{5x} =$

31) $\dfrac{x-2}{7x-12} \div \dfrac{x}{x+3} =$

32) $\dfrac{5x}{x-10} \div \dfrac{5x}{x-5} =$

33) $\dfrac{x^2+10x+16}{x^2+6x+8} \div \dfrac{1}{x+4} =$

34) $\dfrac{x^2-2x-15}{8x+20} \div \dfrac{2}{4x+10} =$

35) $\dfrac{x-4}{x^2-2x-8} \div \dfrac{1}{x-5} =$

## ✎ Simplify each expression.

36) $\dfrac{2}{6x+10} + \dfrac{x-6}{6x+10} =$

37) $\dfrac{4}{x+1} - \dfrac{2}{x+2} =$

38) $\dfrac{2x}{5x+4} + \dfrac{6x}{2x+3} =$

39) $\dfrac{4x}{x+2} + \dfrac{x-3}{x+1} =$

40) $\dfrac{x}{3x+2} + \dfrac{3x}{2x+3} =$

41) $\dfrac{x+5}{4x^2+20x} - \dfrac{x-5}{4x^2+20x} =$

42) $\dfrac{2}{x^2-5x+4} + \dfrac{2}{x^2-4} =$

43) $\dfrac{x-7}{x^2-16} - \dfrac{x-1}{16-x^2} =$

**Effortless Math Education**

✎ **Solve each equation. Remember to check for extraneous solutions.**

44) $\dfrac{2x-3}{x+1} = \dfrac{x+6}{x-2}$

48) $\dfrac{x-2}{x+3} - 1 = \dfrac{1}{x+2}$

45) $\dfrac{3x-2}{9x+1} = \dfrac{2x-5}{6x-5}$

49) $\dfrac{1}{6x^2} = \dfrac{1}{3x^2} - \dfrac{1}{x}$

46) $\dfrac{1}{n-8} - 1 = \dfrac{7}{n-8}$

50) $\dfrac{x+5}{x^2-x} = \dfrac{1}{x^2-x} - \dfrac{x-6}{x-1}$

47) $\dfrac{x+5}{x^2-2x} - 1 = \dfrac{1}{x^2-2x}$

51) $1 = \dfrac{1}{x^2-2x} + \dfrac{x-1}{x}$

✎ **Simplify each expression.**

52) $\dfrac{\frac{12}{3}}{\frac{2}{15}} =$

56) $\dfrac{\frac{12}{x-1}}{\frac{12}{5} - \frac{12}{25}} =$

53) $\dfrac{8}{\frac{8}{x} + \frac{2}{3x}} =$

57) $\dfrac{1 + \frac{2}{x-4}}{1 - \frac{6}{x-4}} =$

54) $\dfrac{x}{\frac{2}{5} - \frac{2}{x}} =$

58) $\dfrac{\frac{x+6}{4}}{\frac{x^2}{2} - \frac{5}{2}} =$

55) $\dfrac{\frac{2}{x+2}}{\frac{8}{x^2+6x+8}} =$

59) $\dfrac{\frac{x-2}{x-6}}{\frac{8}{x-2} + \frac{2}{9}} =$

✎ **Find the maximum and minimum points of the function.**

60) $f(x) = x^3 - 3x + 2$

62) $f(x) = 4x^2 - 3$

61) $f(x) = 3x^2 + 4x + 3$

63) $f(x) = x^3 + x^2 - 8x - 6$

✎ **Solve.**

64) $\frac{x^2-9}{x+3} < 0$

65) $\frac{(x+3)(x+5)}{x+2} \geq 0$

66) $\frac{x+5}{x-4} \geq 0$

67) $\frac{x^2-2x}{x-2} \leq 4$

✎ **What is the domain of the functions?**

68) $y = \frac{x^2-\sqrt{x}}{\sqrt{x}-1}$

69) $y = \frac{x-1}{\sqrt{x^2+2}-1}$

70) $y = -\frac{3}{4}\sqrt{x-1}+5$

✎ **Solve.**

71) When $x$ is 7 and $y$ is 4, find the constant of variation and an equation that inversely relates $y$ and $x$.

72) Refer to the figure on the right. Write an equation to represent the area $S$, Identify the type of variation and the constant of variation.

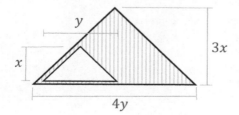

Effortless
Math
Education

# Chapter 11: Answers

1) Rational

2) Irrational

3) Rational

4) Rational

5) Irrational

6) Rational

7) Irrational

8) Irrational

9) $\frac{4}{5}$

10) $\frac{8x^2}{3}$

11) $\frac{5x^2}{3}$

12) $\frac{8}{x-1}$

13) $\frac{5x-1}{8}$

14) $\frac{x+4}{7}$

15) $x - 5$

16) $\frac{x-7}{x-4}$

17) $\frac{x+2}{x-7}$

18)

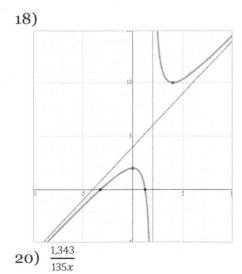

19)

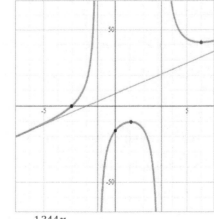

20) $\frac{1,343}{135x}$

21) $\frac{80}{57x}$

22) $\frac{1,344x}{95}$

23) $\frac{2,438x^2}{1,333}$

**Effortless Math Education**

24) $\frac{62}{21x}$

25) $x - 2$

26) $10$

27) $\frac{10}{x+10}$

28) $\frac{32}{5}x$

29) $\frac{40x}{7}$

30) $\frac{x+5}{x-2}$

31) $\frac{(x-2)(x+3)}{x(7x-12)}$

32) $\frac{(x-5)}{(x-10)}$

33) $x + 8$

34) $\frac{(x+3)(x-5)}{4}$

35) $\frac{x-5}{x+2}$

36) $\frac{-4+x}{6x+10}$

37) $\frac{2x+6}{(x+1)(x+2)}$

38) $\frac{34x^2+30x}{(5x+4)(2x+3)}$

39) $\frac{5x^2+3x-6}{(x+2)(x+1)}$

40) $\frac{11x^2+9x}{(3x+2)(2x+3)}$

41) $\frac{5}{2x(x+5)}$

42) $\frac{4x^2-10x}{(x-1)(x-4)(x+2)(x-2)}$

43) $\frac{2}{x+4}$

44) $\{0, 14\}$

45) $\{-\frac{15}{16}\}$

46) $\{2\}$

47) $\{4, -1\}$

48) $\{-\frac{13}{6}\}$

49) $\{\frac{1}{6}\}$

50) $\{4\}$

51) $\{3\}$

52) $30$

53) $\frac{12x}{13}$

**Effortless Math Education**

54) $\frac{5x^2}{2x-10}$

55) $\frac{(x+4)}{4}$

56) $\frac{25}{4(x-1)}$

57) $\frac{x-2}{x-10}$

58) $\frac{x+6}{2x^2-10}$

59) $\frac{9(x-2)^2}{(2x+68)(x-6)}$

60) Maximum: $(-1, 4)$

   Minimum: $(1, 0)$

61) Minimum: $\left(-\frac{2}{3}, \frac{5}{3}\right)$

62) Minimum: $(0, -3)$

63) Maximum: $(-2, 6)$

   Minimum: $\left(\frac{4}{3}, -\frac{338}{27}\right)$

64) $(-\infty, -3) \cup (-3, 3)$

65) $[-5, -3] \cup (-2, +\infty)$

66) $(-\infty, -5] \cup (4, +\infty)$

67) $(-\infty, 2) \cup (2, 4]$

68) Domain: $[0, 1) \cup (1, +\infty)$

69) Domain: $(-\infty, +\infty)$

70) Domain: $[1, +\infty)$

71) $y = \frac{28}{x}$

72) The area varies jointly as $x$ and $y$. The constant of variation is $\frac{11}{2}$.

**Effortless**
**Math**
**Education**

# CHAPTER

## 12 Conics

Math topics that you'll learn in this chapter:

- ☑ Equation of a Parabola
- ☑ Finding the Focus, Vertex, and Directrix of a Parabola
- ☑ Standard From of a Circle
- ☑ Finding the Center and the Radius of Circles
- ☑ Equation of Ellipse
- ☑ Hyperbola in Standard Form
- ☑ Classifying a Conic Section (in Standard Form)

161

# Equation of a Parabola

- The standard form of a parabola:

When it opens up or down:

$(x - h)^2 = 4p(y - k)$, Vertex: $(h, k)$, Directrix: $y = k - p$, Focus: $(h, k + p)$.

When it opens right or left:

$(y - k)^2 = 4p(x - h)$, Vertex: $(h, k)$, Directrix: $x = h - p$, Focus: $(h + p, k)$.

## Examples:

**Example 1.** Write the equation of the parabola: Vertex: (0,0) and Focus: (0,4).

*Solution*: The standard form of a parabola:

$(x - h)^2 = 4p(y - k)$.

Vertex: $(h, k) = (0,0)$, then: $h = 0$, $k = 0$.

Focus: $(h, k + p) = (0,4)$, then:

$k + p = 4 \rightarrow p = 4$.

Put in formula: $(x - 0)^2 = 4(4)(y - 0)$, then:

$x^2 = 16y$.

**Example 2.** Write the equation of the parabola: Vertex: (2,3) and Focus: (2,4).

*Solution*: The standard form of a parabola:

$(x - h)^2 = 4p(y - k)$.

Vertex: $(h, k) = (2,3)$, then: $h = 2$, $k = 3$.

Focus: $(h, k + p) = (2,4)$, then:

$k + p = 4 \rightarrow 3 + p = 4 \rightarrow p = 1$.

Put in the standard form: $(x - 2)^2 = 4(1)(y - 3)$, then:

$$(x - 2)^2 = 4(y - 3).$$

# Finding the Focus, Vertex, and Directrix of a Parabola

- The standard form of a parabola:

  When it opens up or down:

  $(x - h)^2 = 4p(y - k)$, Vertex: $(h, k)$, Directrix: $y = k - p$, Focus: $(h, k + p)$.

  When it opens right or left:

  $(y - k)^2 = 4p(x - h)$, Vertex: $(h, k)$, Directrix: $x = h - p$, Focus: $(h + p, k)$.

- The $x$ value of the vertex of an up-down facing parabola of the form:

  $y = ax^2 + bx + c$ is: $x_v = -\frac{b}{2a}$.

## Examples:

**Example 1.** Find the vertex of parabola. $y = x^2 + 4x$

*Solution*: The parabola params are: $a = 1$, $b = 4$ and $c = 0$.

$x_v = -\frac{b}{2a} \rightarrow x_v = -\frac{4}{2(1)} = -2$.

Use $x_v$ to find $y_v$: $y_v = ax^2 + bx + c$, then:

$y_v = (1)(-2)^2 + (4)(-2) + 0 \rightarrow y_v = -4$.

Therefore, the vertex of the parabola is: $(-2, -4)$.

**Example 2.** Write the vertex form equation of a parabola. $y = x^2 - 6x + 5$

*Solution*: First, evaluate the $x$ value of the vertex from the formula $x_v = -\frac{b}{2a}$. Then: $x_v = -\frac{(-6)}{2(1)} = 3$. To find the $y$ value of the vertex, substitute $x_v = 3$ in the equation: $y_v = (3)^2 - 6(3) + 5 = -4$. Now, plug the vertex $(x_v, y_v) = (3, -4)$ into the vertex formula of parabola and simplify:

$(x - 3)^2 = 4p(y - (-4)) \rightarrow x^2 - 6x + 9 = 4p(y + 4) \rightarrow x^2 - 6x + 9 = 4py + 16p$.

Rewrite the obtained equation as $4py = x^2 - 6x + (9 - 16p)$. Then, compare it's with the given equation and obtain the value of $p$. Therefore:

$4p = 1 \rightarrow p = \frac{1}{4}$. Finally,

$p = \frac{1}{4} \rightarrow (x - 3)^2 = 4\left(\frac{1}{4}\right)(y + 4) \rightarrow (x - 3)^2 = (y + 4)$.

# Standard Form of a Circle

- Equation of circles in standard form: $(x - h)^2 + (y - k)^2 = r^2$.

- Center: $(h, k)$, Radius: $r$.

- General format: $ax^2 + by^2 + cx + dy + e = 0$.

## Examples:

**Write the standard form equation of each circle.**

**Example 1.** Center: $(-9, -12)$, Radius: 4.

***Solution***: $(x - h)^2 + (y - k)^2 = r^2$ is the circle equation with a radius $r$, centered at $(h, k)$. We have:

$h = -9$, $k = -12$ and $r = 4$.

Then:

$$\left(x - (-9)\right)^2 + \left(y - (-12)\right)^2 = (4)^2 \rightarrow (x + 9)^2 + (y + 12)^2 = 16.$$

**Example 2.** $x^2 + y^2 - 8x - 6y + 21 = 0$.

***Solution***: $(x - h)^2 + (y - k)^2 = r^2$ is the circle equation with a radius $r$, centered at $(h, k)$. First move the loose number to the right side:

$x^2 + y^2 - 8x - 6y = -21$.

Group $x$ −variables and $y$ −variables together:

$(x^2 - 8x) + (y^2 - 6y) = -21$.

Convert $x$ to square form:

$(x^2 - 8x + 16) + (y^2 - 6y) = -21 + 16 \rightarrow (x - 4)^2 + (y^2 - 6y) = -5$.

Convert $y$ to square form:

$(x - 4)^2 + (y^2 - 6y + 9) = -21 + 16 + 9 \rightarrow (x - 4)^2 + (y - 3)^2 = 4$.

Then:

$$(x - 4)^2 + (y - 3)^2 = 2^2.$$

# Finding the Center and the Radius of Circles

- $(x - h)^2 + (y - k)^2 = r^2$

  Center: $(h, k)$, Radius: $r$.

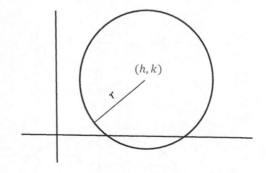

## Examples:

**Identify the center and radius.**

**Example 1.** $x^2 + y^2 - 4y + 3 = 0$.

**Solution**: $(x - h)^2 + (y - k)^2 = r^2$ is the circle equation with a radius $r$, centered at $(h, k)$. Rewrite $x^2 + y^2 - 4y + 3 = 0$ in the form of the standard circle equation:

Group $x$ −variables and $y$ −variables together:

$(x^2) + (y^2 - 4y) = -3$.

Convert $x$ and $y$ to square form:

$(x^2) + (y^2 - 4y + 4) - 4 = -3$.

Therefore:

$(x - 0)^2 + (y - 2)^2 = 1^2$.

Then:

Center: $(0, 2)$ and $r = 1$.

**Example 2.** $4x + x^2 - 6y = 24 - y^2$.

**Solution**: $(x - h)^2 + (y - k)^2 = r^2$ is the circle equation with a radius $r$, centered at $(h, k)$. Rewrite $4x + x^2 - 6y = 24 - y^2$ in the form of the standard circle equation:

$\left(x - (-2)\right)^2 + (y - 3)^2 = \left(\sqrt{37}\right)^2$.

Then: Center: $(-2, 3)$ and $r = \sqrt{37}$.

# Equation of an Ellipse

- Horizontal: $\frac{(x-h)^2}{a^2} + \frac{(y-k)^2}{b^2} = 1$     - Vertical: $\frac{(x-h)^2}{b^2} + \frac{(y-k)^2}{a^2} = 1$

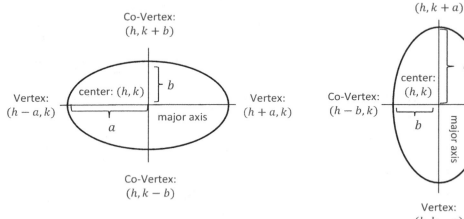

Horizontal fosi: $(h+c,k), (h-c,k)$

Vertical fosi: $(h,k+c), (,kh-c)$

- Horizontal Ellipse Vertices: the vertices are the two points on the ellipse that intersect the major axis for an ellipse with the major axis parallel to the $x$ −axis, the vertices are: $(h+a,k), (h-a,k)$

- Horizontal Ellipse Foci: $(h+c,k), (h-c,k)$, where $c = \sqrt{a^2 - b^2}$ is the distance from the center $(h,k)$ to a focus.

## Example:

Find vertices of $\frac{x^2}{169} + \frac{y^2}{64} = 1$.

**Solution**: Rewrite $\frac{x^2}{169} + \frac{y^2}{64} = 1$ in the form of the standard ellipse equation:
$\frac{(x-h)^2}{a^2} + \frac{(y-k)^2}{b^2} = 1 \to \frac{(x-0)^2}{13^2} + \frac{(y-0)^2}{8^2} = 1$.
Then:
$(h,k) = (0,0)$, $a = 13$ and $b = 8$.
To find the vertices, we have:
$$(h+a,k) \to (0+13,0) = (13,0),$$
$$(h-a,k) \to (0-13,0) = (-13.0).$$

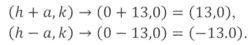

# Hyperbola in Standard Form

Up/down Hyperbola:

$$\frac{(y-k)^2}{a^2} - \frac{(x-h)^2}{b^2} = 1$$

center: $(h, k)$

foci: $(h, k \pm c)$

$c^2 = a^2 + b^2$

vertices: $(h, k \pm a)$

transverse axis: $x = h$

(Parallel to $y$ −axis)

asymptotes: $y - k = \pm \frac{a}{b}(x - h)$

Left/right Hyperbola:

$$\frac{(x-h)^2}{a^2} - \frac{(y-k)^2}{b^2} = 1$$

center: $(h, k)$

foci: $(h \pm c, k)$

$c^2 = a^2 + b^2$

vertices: $(h \pm a, k)$

transverse axis: $y = k$

(Parallel to $x$ −axis)

asymptotes: $y - k = \pm \frac{b}{a}(x - h)$

## Example:

Find the center and foci of $-x^2 + y^2 - 18x - 14y - 132 = 0$.

*Solution:* Rewrite in standard form. Add 132 to both sides:

$-x^2 + y^2 - 18x - 14y = 132$.

Factor out the coefficient of square terms:

$(y^2 - 14y) - (x^2 + 18x) = 132$.

Convert $x$ and $y$ to square form:

$(y^2 - 14y + 49) - (x^2 + 18x + 81) = 132 - 81 + 49 \rightarrow (y - 7)^2 - (x + 9)^2 = 100$.

Divide by 100: $-\frac{(x+9)^2}{100} + \frac{(y-7)^2}{100} = 1 \rightarrow \frac{(y-7)^2}{10^2} - \frac{(x-(-9))^2}{10^2} = 1$.

Then: $(h, k) = (-9, 7)$, $a = 10$, $b = 10$ and center is: $(-9, 7)$.

Foci: $(-9, 7 + c)$, $(-9, 7 - c)$.

Compute $c$: $c = \sqrt{10^2 + 10^2} = 10\sqrt{2}$.

Then: $(-9, 7 + 10\sqrt{2})$, $(-9, 7 - 10\sqrt{2})$.

# Classifying a Conic Section (in Standard Form)

| Conic section | Standard form of equation | | |
|---|---|---|---|
| Parabola | $y = a(x - h)^2 + k$ | $x = a(y - k)^2 + h$ | $a = 4p$ |
| Circle | $(x - h)^2 + (y - k)^2 = r^2$ | | |
| Ellipse | $\frac{(x-h)^2}{a^2} + \frac{(y-k)^2}{b^2} = 1$ | $\frac{(y-k)^2}{a^2} + \frac{(x-h)^2}{b^2} = 1$ | |
| Hyperbola | $\frac{(x-h)^2}{a^2} - \frac{(y-k)^2}{b^2} = 1$ | $\frac{(y-k)^2}{a^2} - \frac{(x-h)^2}{b^2} = 1$ | |

## Examples:

**Example 1.** Write this equation in standard form. $-x^2 + 8x + y - 17 = 0$.

*Solution*: It's a parabola. Rewrite in standard form: $-x^2 + 8x + y - 17 = 0$.

Rewrite as: $y = x^2 - 8x + 17$, complete the square $x^2 - 8x + 17 = (x - 4)^2 + 1$.

Then: $y = (x - 4)^2 + 1$.

Subtract 1 from both sides: $y - 1 = (x - 4)^2$.

Rewrite in standard form: $4 \cdot \frac{1}{4}(y - 1) = (x - 4)^2$.

**Example 2.** Write this equation in standard form. $x^2 - 4y^2 + 6x - 8y + 1 = 0$.

*Solution*: It's a hyperbola. First subtract 1 from both sides:

$x^2 - 4y^2 + 6x - 8y + 1 = 0 \rightarrow x^2 - 4y^2 + 6x - 8y = -1$.

Factor out the coefficient of square terms: $(x^2 + 6x) - 4(y^2 + 2y) = -1$.

Divide by the coefficient of square terms: $\frac{1}{4}(x^2 + 6x) - (y^2 + 2y) = -\frac{1}{4}$.

Convert $x$ and $y$ to square form: $\frac{1}{4}(x^2 + 6x + 9) - (y^2 + 2y + 1) = -\frac{1}{4} + \frac{1}{4}(9) - 1$.

Convert $y$ to square form: $\frac{1}{4}(x + 3)^2 - (y + 1)^2 = -\frac{1}{4} + \frac{1}{4}(9) - 1$.

Then: $\frac{1}{4}(x + 3)^2 - (y + 1)^2 = 1 \rightarrow \frac{(x+3)^2}{4} - \frac{(y+1)^2}{1} = 1 \rightarrow \frac{(x-(-3))^2}{2^2} - \frac{(y-(-1))^2}{1^2} = 1$.

# Chapter 12: Practices

✏️ **Write the equation of the following parabolas.**

1) Vertex $(0, 0)$ and Focus $(0, 2)$

5) Vertex $(2, 2)$ and Focus $(2, 6)$

2) Vertex $(3, 2)$ and Focus $(3, 4)$

6) Vertex $(0, 1)$ and Focus $(0, 2)$

3) Vertex $(1, 1)$ and Focus $(1, 6)$

7) Vertex $(2, 1)$ and Focus $(4, 1)$

4) Vertex $(-1, 2)$ and Focus $(-1, 5)$

8) Vertex $(5, 0)$ and Focus $(9, 0)$

✏️ **Write the vertex form equation of each parabola.**

9) $y = x^2 + 8x$

11) $y + 6 = (x + 3)^2$

10) $y = x^2 - 6x + 5$

12) $y = x^2 + 10x + 33$

✏️ **Write the standard form equation of each circle.**

13) $y^2 + 2x + x^2 = 24y - 120$

14) $x^2 + y^2 - 2y - 15 = 0$

15) $8x + x^2 - 2y = 64 - y^2$

16) Center: $(-5, -6)$, Radius: 9

17) Center: $(-12, -5)$, Area: $4\pi$

18) Center: $(-11, -14)$, Area: $16\pi$

19) Center: $(-3, 2)$, Circumference: $2\pi$

20) Center: $(15, 14)$, Circumference: $2\pi\sqrt{15}$

Effortless
Math
Education

✍ **Identify the center and radius of each.**

21) $(x - 2)^2 + (y + 5)^2 = 10$        23) $(x - 2)^2 + (y + 6)^2 = 9$

22) $x^2 + (y - 1)^2 = 4$        24) $(x + 14)^2 + (y - 5)^2 = 16$

✍ **Identify the vertices, co-vertices, and foci.**

25) $\frac{x^2}{36} + \frac{y^2}{16} = 1$        28) $\frac{(x-3)^2}{49} + \frac{(y-9)^2}{4} = 1$

26) $\frac{x^2}{49} + \frac{y^2}{169} = 1$        29) $\frac{x^2}{64} + \frac{(y-8)^2}{9} = 1$

27) $\frac{(x+5)^2}{81} + \frac{(y-1)^2}{144} = 1$        30) $\frac{x^2}{64} + \frac{(y-6)^2}{121} = 1$

✍ **Identify the vertices, foci, and direction of the opening of each.**

31) $\frac{y^2}{25} - \frac{x^2}{16} = 1$        34) $\frac{x^2}{81} - \frac{y^2}{4} = 1$

32) $\frac{x^2}{121} - \frac{y^2}{36} = 1$        35) $\frac{(x+2)^2}{169} - \frac{(y+8)^2}{4} = 1$

33) $\frac{x^2}{121} - \frac{y^2}{81} = 1$        36) $\frac{(y+8)^2}{36} - \frac{(x+2)^2}{25} = 1$

✍ **Classify each conic section and write its equation in standard form.**

37) $3x^2 + 30x + y + 79 = 0$        40) $-9x^2 + y^2 - 72x - 153 = 0$

38) $x^2 + y^2 + 4x - 2y - 18 = 0$        41) $-2y^2 + x - 20y - 49 = 0$

39) $49x^2 + 9y^2 + 392x + 343 = 0$        42) $-x^2 + 10x + y - 21 = 0$

# Chapter 12: Answers

1) $x^2 = 8y$

2) $(x-3)^2 = 8(y-2)$

3) $(x-1)^2 = 20(y-1)$

4) $(x+1)^2 = 12(y-2)$

5) $(x-2)^2 = 16(y-2)$

6) $x^2 = 4(y-1)$

7) $(y-1)^2 = 8(x-2)$

8) $y^2 = 16(x-5)$

9) $y = (x+4)^2 - 16$

10) $y = (x-3)^2 - 4$

11) $y = (x+3)^2 - 6$

12) $y = (x+5)^2 + 8$

13) $(x+1)^2 + (y-12)^2 = 25$

14) $x^2 + (y-1)^2 = 16$

15) $(x+4)^2 + (y-1)^2 = 81$

16) $(x+5)^2 + (y+6)^2 = 81$

17) $(x+12)^2 + (y+5)^2 = 4$

18) $(x+11)^2 + (y+14)^2 = 16$

19) $(x+3)^2 + (y-2)^2 = 1$

20) $(x-15)^2 + (y-14)^2 = 15$

21) Center: $(2,-5)$, Radius: $\sqrt{10}$

22) Center: $(0,1)$, Radius: 2

23) Center: $(2,-6)$, Radius: 3

24) Center: $(-14,5)$, Radius: 4

25) Vertices: $(6,0)$, $(-6,0)$

   Co–vertices: $(0,4)$, $(0,-4)$

   Foci: $(2\sqrt{5},0)$, $(-2\sqrt{5},0)$

26) Vertices: $(0,13)$, $(0,-13)$

   Co–vertices: $(7,0)$, $(-7,0)$

   Foci: $(0,2\sqrt{30})$, $(0,-2\sqrt{30})$

27) Vertices: $(-5,13)$, $(-5,-11)$

   Co–vertices: $(4,1)$, $(-14,1)$

   Foci: $(-5,1 \pm 3\sqrt{7})$

28) Vertices: $(10,9)$, $(-4,9)$

   Co–vertices: $(3,11)$, $(3,7)$

   Foci: $(3 \pm 3\sqrt{5},9)$

**Effortless Math Education**

29) Vertices: $(8,8)$, $(-8,8)$

   Co–vertices: $(0,11)$, $(0,5)$

   Foci: $(\sqrt{55}, 8)$, $(-\sqrt{55}, 8)$

30) Vertices: $(0,17)$, $(0,-5)$

   Co–vertices: $(8,6)$, $(-8,6)$

   Foci: $(0, 6 \pm \sqrt{57})$

31) Vertices: $(0,5)$, $(0,-5)$

   Foci: $(0, \sqrt{41})$, $(0, -\sqrt{41})$

   Opens up/down

32) Vertices: $(11,0)$, $(-11,0)$

   Foci: $(\sqrt{157}, 0)$, $(-\sqrt{157}, 0)$

   Opens left/right

33) Vertices: $(11,0)$, $(-11,0)$

   Foci: $(\sqrt{202}, 0)$, $(-\sqrt{202}, 0)$

   Opens left/right

34) Vertices: $(9,0)$, $(-9,0)$

   Foci: $(\sqrt{85}, 0)$, $(-\sqrt{85}, 0)$

   Opens left/right

35) Vertices: $(11, -8)$, $(-15, -8)$

   Foci: $(-2 \pm \sqrt{173}, -8)$

   Opens left/right

36) Vertices: $(-2, -2)$, $(-2, -14)$

   Foci: $(-2, -8 \pm \sqrt{61})$

   Opens up/down

37) Parabola, $4(-\frac{1}{12})(y - (-4)) = (x - (-5))^2$

38) Circle, $(x - (-2))^2 + (y - 1)^2 = (\sqrt{23})^2$

39) Ellipse, $\dfrac{(x-(-4))^2}{3^2} + \dfrac{(y)^2}{7^2} = 1$

40) Hyperbola, $\dfrac{(y)^2}{3^2} - \dfrac{(x-(-4))^2}{1^2} = 1$

41) Parabola, $(4)\left(\frac{1}{8}\right)(x - (-1)) = (y - (-5))^2$

42) Parabola, $(4)\left(\frac{1}{4}\right)(y - (-4)) = (x - 5)^2$

# CHAPTER

## 13 Sequences and Series

Math topics that you'll learn in this chapter:

- ☑ Arithmetic Sequences
- ☑ Geometric Sequences
- ☑ Arithmetic Series
- ☑ Finite Geometric Series
- ☑ Infinite Geometric Series
- ☑ Pascal's Triangle
- ☑ Binomial Theorem
- ☑ Sigma Notation (Summation Notation)
- ☑ Alternate Series

173

# Arithmetic Sequences

- A sequence of numbers such that the difference between the consecutive terms is constant is called an arithmetic sequence. For example, the sequence 6, 8, 10, 12, 14, ⋯ is an arithmetic sequence with common difference of 2.

- To find any term in an arithmetic sequence use this formula:

$$x_n = a + d(n - 1)$$

$a$ = the first term, $d$ = the common difference between terms, $n$ = number of items.

## Examples:

**Example 1.** Find the first five terms of the sequence. $a_8 = 38$, $d = 3$

*Solution*: First, we need to find $a_1$ or $a$.

Use the arithmetic sequence formula:

$x_n = a + d(n - 1)$.

If $a_8 = 38$, then $n = 8$. Rewrite the formula and put the values provided:

$x_n = a + d(n - 1) \rightarrow 38 = a + 3(8 - 1) = a + 21$.

Now solve for $a$:

$38 = a + 21 \rightarrow a = 38 - 21 = 17$.

First five terms:

17, 20, 23, 26, 29.

**Example 2.** Given the first term and the common difference of an arithmetic sequence find the first five terms. $a_1 = 18$, $d = 2$.

*Solution*: Use the arithmetic sequence formula:

$x_n = a + d(n - 1)$.

First five terms:

18, 20, 22, 24, 26.

# Geometric Sequences

- It is a sequence of numbers where each term after the first is found by multiplying the previous item by the common ratio, a fixed, non-zero number. For example, the sequence 2, 4, 8, 16, 32, ⋯ is a geometric sequence with a common ratio of 2.

- To find any term in a geometric sequence use this formula: $x_n = ar^{(n-1)}$.

  $a$ = the first term, $r$ = the common ratio, $n$ = number of items.

## Examples:

**Example 1.** Given the first term and the common ratio of a geometric sequence find the first five terms of the sequence. $a_1 = 3, r = -2$

**Solution**: Use geometric sequence formula:

$x_n = ar^{(n-1)} \rightarrow x_n = 3(-2)^{n-1}$.

If $n = 1$, then: $x_1 = 3(-2)^{1-1} = 3(1) = 3$.

First five terms:

$3, -6, 12, -24, 48$.

**Example 2.** Given two terms in a geometric sequence find the 8th term. $a_3 = 10$, and $a_5 = 40$.

**Solution**: Use geometric sequence formula:

$x_n = ar^{(n-1)} \rightarrow a_3 = ar^{(3-1)} = ar^2 = 10$.

$x_n = ar^{(n-1)} \rightarrow a_5 = ar^{(5-1)} = ar^4 = 40$.

Now divide $a_5$ by $a_3$. Then: $\frac{a_5}{a_3} = \frac{ar^4}{ar^2} = \frac{40}{10}$.

Now simplifies:

$\frac{ar^4}{ar^2} = \frac{40}{10} \rightarrow r^2 = 4 \rightarrow r = 2$.

We can find $a$ now: $ar^2 = 10 \rightarrow a(2^2) = 10 \rightarrow a = 2.5$.

Use the formula to find the 8th term:

$x_n = ar^{(n-1)} \rightarrow a_8 = (2.5)(2)^{(8-1)} = 2.5(128) = 320$.

bit.ly/3pxGzkP

Find more at

# Arithmetic Series

- An arithmetic series is the sum of sequence in which each term is computed from the previous one by adding (or subtracting) a constant value $d$.

- The sum of the sequence of the first $n$ terms is given by:

$$S_n = \sum_{k=1}^{n} a_k = \sum_{k=1}^{n} [a_1 + (k-1)d] = na_1 + d \sum_{k=1}^{n} (k-1)$$

- Using the sum identify $\sum_{k=1}^{n} \frac{1}{2} n(n+1)$, then gives:

$$S_n = na_1 + \frac{1}{2} dn(n-1) = \frac{1}{2} n[2a_1 + d(n-1)]$$

- Note that: $a_1 + a_n = a_1 + [a_1 + d(n-1)] = 2a_1 + d(n-1)$, so:

$$S_n = \frac{1}{2} n(a_1 + a_n)$$

## Examples:

**Example 1.** In the arithmetic series 4, 11, 18, $\cdots$ find the sum of the first 10 terms.

***Solution***: $a_1 = 4$, $d = 11 - 4 = 7$, $n = 10$.

Use the arithmetic series formula to find the sum:

$S_n = \frac{1}{2} n[2a_1 + d(n-1)]$.

Therefore:

$S_{10} = \frac{10}{2}[2(4) + 7(10-1)] \rightarrow S_{10} = 5(8 + 63) \rightarrow S_{10} = 355$.

**Example 2.** Find the sum of the first four terms of the sequence. $a_{10} = 46$, $d = 4$.

***Solution***: First, we need to find $a_1$ or $a$. Use the arithmetic sequence formula:

$a_n = a + d(n-1)$.

If $a_{10} = 46$, then $n = 10$. Rewrite the formula and put the values provided: $a_n = a + d(n-1) \rightarrow 46 = a + 4(10-1) = a + 36$, now solve for $a$.

$46 = a + 36 \rightarrow a = 46 - 36 = 10$.

First four terms: 10, 14, 18, 22. Therefore:

$S_n = \frac{1}{2} n[2a_1 + d(n-1)] = \frac{1}{2}(4)[2(10) + 4(4-1)] = 64$.

# Finite Geometric Series

- The sum of a geometric series is finite when the absolute value of the ratio is less than 1.

- Finite Geometric Series Formula: $S_n = \sum\limits_{i=1}^{n} a_1 r^{i-1} = a_1 \left( \frac{1-r^n}{1-r} \right)$.

## Examples:

**Evaluate each geometric series described.**

**Example 1.** $\sum\limits_{n=1}^{4} 3^{n-1}$.

***Solution***: Use this formula:

$$S_n = \sum\limits_{i=1}^{n} a_1 r^{i-1} = a_1 \left( \frac{1-r^n}{1-r} \right) \rightarrow \sum\limits_{n=1}^{4} 3^{n-1} = (1) \left( \frac{1-3^4}{1-3} \right).$$

Then:

$(1) \left( \frac{1-3^4}{1-3} \right) = 1 \left( \frac{1-81}{1-3} \right) = \left( \frac{-80}{-2} \right) 1 = 40$.

**Example 2.** $\sum\limits_{n=1}^{5} -2^{n-1}$.

***Solution***: Use this formula:

$$S_n = \sum\limits_{i=1}^{n} a_1 r^{i-1} = a_1 \left( \frac{1-r^n}{1-r} \right) \rightarrow \sum\limits_{n=1}^{5} -2^{n-1} = (-1) \left( \frac{1-2^5}{1-2} \right).$$

Then: $(-1) \left( \frac{1-32}{1-2} \right) = (-1) \left( \frac{-31}{-1} \right) = -31$.

**Example 3.** $\sum\limits_{n=1}^{7} \left( -\frac{1}{2} \right)^{n-1}$.

***Solution***: Use this formula:

$$S_n = \sum\limits_{i=1}^{n} a_1 r^{i-1} = a_1 \left( \frac{1-r^n}{1-r} \right) \rightarrow \sum\limits_{n=1}^{7} \left( -\frac{1}{2} \right)^{n-1} = (1) \left( \frac{1-\left(-\frac{1}{2}\right)^7}{1-\left(-\frac{1}{2}\right)} \right).$$

Then:

$(1) \left( \frac{1+\frac{1}{128}}{1+\frac{1}{2}} \right) = \left( \frac{\frac{129}{128}}{\frac{3}{2}} \right) = \frac{129}{192} = \frac{43}{64}$.

# Infinite Geometric Series

- Infinite Geometric Series: The sum of a geometric series is infinite when the absolute value of the ratio is more than 1.

- Infinite Geometric Series Formula: $S = \sum_{i=0}^{\infty} a_i r^i = \frac{a_1}{1-r}$.

## Examples:

**Example 1.** Evaluate the infinite geometric series described. $\sum_{i=1}^{\infty} (-\frac{2}{3})^{i-1}$

*Solution*: Since the absolute value of the ratio is $\frac{2}{3}$ and less than 1, the sum of a geometric series is finite.

Therefore, by using this formula:

$$\sum_{i=0}^{\infty} a_i r^i = \frac{a_1}{1-r} \to \sum_{i=1}^{\infty} (-\frac{2}{3})^{i-1} = \frac{1}{1-(-\frac{2}{3})} = \frac{1}{\frac{5}{3}} = \frac{3}{5}.$$

**Example 2.** Evaluate the infinite geometric series described. $\sum_{i=1}^{\infty} (\frac{1}{3})^{i-1}$

*Solution*: The absolute value of the ratio is $\frac{1}{3}$. Use this formula:

$$\sum_{i=0}^{\infty} a_i r^i = \frac{a_1}{1-r} \to \sum_{i=1}^{\infty} (\frac{1}{3})^{i-1} = \frac{1}{1-\frac{1}{3}} = \frac{1}{\frac{2}{3}} = \frac{3}{2}.$$

**Example 3.** Evaluate the infinite geometric series described. $\sum_{k=1}^{\infty} -2(\frac{1}{4})^{k-1}$

*Solution*: The absolute value of the ratio is $\frac{1}{4}$. Use this formula:

$$\sum_{i=0}^{\infty} a_i r^i = \frac{a_1}{1-r}.$$

Put $a_1 = -2$ and $r = \frac{1}{4}$. Therefore:

$$\sum_{i=0}^{\infty} a_i r^i = \frac{a_1}{1-r} \to \sum_{k=1}^{\infty} -2\left(\frac{1}{4}\right)^{k-1} = \frac{-2}{1-\frac{1}{4}} = \frac{-2}{\frac{3}{4}} = \frac{-8}{3}.$$

**Example 4.** Evaluate the infinite geometric series described. $\sum_{k=1}^{\infty} (\frac{1}{4})7^{k-1}$

*Solution*: Since the absolute value of the ratio is 7 and more than 1, the sum of a geometric series is infinite.

# Pascal's Triangle

- Pascal's triangle is shown below:

$$_0C_0$$

$$_1C_0 \quad _1C_1$$

$$_2C_0 \quad _2C_1 \quad _2C_2$$

$$_3C_0 \quad _3C_1 \quad _3C_2 \quad _3C_3$$

$$\cdot\cdot \qquad \vdots \qquad \cdot\cdot$$

$$_nC_0 \quad _nC_1 \quad _nC_2 \quad \cdots \quad _nC_{n-2} \quad _nC_{n-1} \quad _nC_n$$

$$_{n+1}C_0 \quad _{n+1}C_1 \quad _{n+1}C_2 \quad \cdots \quad _{n+1}C_{n-1} \quad _{n+1}C_n \quad _{n+1}C_{n+1}$$

- The $n$th row of Pascal's triangle contains $n + 1$ components.

- In the nth row of Pascal's triangle, the $k$th component is equal to $_nC_{k-1}$.

- For all entries in Pascal's triangle in row $n$: $\displaystyle\sum_{k=0}^{n} {_nC_k} = 2^n$.

- $_nC_{k-1} + {_nC_k} = {_{n+1}C_k}$, where $0 < k \leq n$.

# Examples:

**Example 1.** Find the 5th entry in row 7 of Pascal's triangle.
**Solution:** The $k$th entry in row $n$ of Pascal's triangle is $_nC_{k-1}$.
(5th entry in row 7) $= {_7C_{5-1}} = {_7C_4} = \frac{7!}{3!4!} = 35$.

**Example 2.** Find the 8th entry in row 10 of Pascal's triangle.
**Solution:** The $k$th entry in row $n$ of Pascal's triangle is $_nC_{k-1}$.
(8th entry in row 10) $= {_{10}C_{8-1}} = {_{10}C_7} = \frac{10!}{7!3!} = 120$.

**Example 3.** Find the location of $_{10}C_2$ entry in Pascal's triangle. Then give its value of it.
**Solution:** We know that the $k$th entry in row $n$ of Pascal's triangle is $_nC_{k-1}$.
Then the value of $n$ and $k$ for $_{10}C_2$ is $n = 10$ and $k - 1 = 2 \to k = 3$. So $_{10}C_2$ is
the 3rd entry in row 10 of Pascal's triangle. Now, $_{10}C_2 = \frac{10!}{2!8!} = 45$.

bit.ly/3Xk1h5X

Find more at

# Binomial Theorem

- The formula of the binomial theorem for positive integer $n$ is as follows:

$$(x + y)^n = \sum_{k=0}^{n} \binom{n}{k} x^{n-k} y^k$$

$$= \binom{n}{0} x^n y^0 + \binom{n}{1} x^{n-1} y^1 + \cdots + \binom{n}{n-1} x^1 y^{n-1} + \binom{n}{n} x^0 y^n$$

Where $\binom{n}{k} = \frac{n!}{k!(n-k)!}$, and $0 \le k \le n$.

- In the expansion of $(x + y)^n$, there are $n + 1$ terms and the $k$th term is equal to:

$$\binom{n}{k-1} x^{n-k+1} y^{k-1}$$

## Examples:

**Example 1.** Write the expansion $(x + y)^4$.

**Solution:** $(x + y)^4 = \sum_{k=0}^{4} \binom{4}{k} x^{4-k} y^k$

$$= \binom{4}{0} x^4 + \binom{4}{1} x^3 y + \binom{4}{2} x^2 y^2 + \binom{4}{3} xy^3 + \binom{4}{4} y^4$$

$$= x^4 + 4x^3 y + 6x^2 y^2 + 4xy^3 + y^4$$

**Example 2.** Write the 3rd term of the expansion of $(a - 1)^5$.
**Solution:** Use this formula: The 3th term $= \binom{n}{k-1} x^{n-k+1} y^{k-1}$

The 3th term $= \binom{5}{3-1} (a)^{5-3+1} (-1)^{3-1}$

$$= \binom{5}{2} (a)^3 (-1)^2$$

$$= 10a^3$$

**Example 3.** Write the expansion $(2b + 2)^3$.

**Solution:** $(2b + 2)^3 = \sum_{k=0}^{3} \binom{3}{k} (2b)^{3-k} (2)^k$

$$= \binom{3}{0} (2b)^3 + \binom{3}{1} (2b)^2 (2) + \binom{3}{2} (2b)(2)^2 + \binom{3}{3} (2)^3$$

$$= 8b^3 + 24b^2 + 24b + 8$$

# Sigma Notation (Summation Notation)

- Summation notation is a way to express the sum of terms of a sequence in abbreviated form.

- Let $a_1, a_2, \cdots, a_i, \cdots$ is a sequence with the starting term $a_1$ and the $i$th term $a_i$. The representation of the sum of the $k$th term to the $n$th term of this sequence is as follows:

Stopping point ←

Summation sign
Called sigma ← $\displaystyle\sum_{i=k}^{n} a_i$ → Formula for general

index ← └ └ → Starting point

- For sequences $a_i$ and $b_i$ and positive integer $n$:

$\circ \displaystyle\sum_{i=1}^{n} ca_i = c\sum_{i=1}^{n} a_i$ 

$\circ \displaystyle\sum_{i=1}^{n} i = \frac{n(n+1)}{2}$

$\circ \displaystyle\sum_{i=1}^{n} (a_i + b_i) = \sum_{i=1}^{n} a_i + \sum_{i=1}^{n} b_i$ 

$\circ \displaystyle\sum_{i=1}^{n} i^2 = \frac{n(n+1)(2n+1)}{6}$

$\circ \displaystyle\sum_{i=1}^{n} c = nc$

## Example:

Evaluate: $\displaystyle\sum_{k=1}^{10} (k^2 - 1)$.

**Solution**: Using this property $\displaystyle\sum_{i=1}^{n} (a_i + b_i) = \sum_{i=1}^{n} a_i + \sum_{i=1}^{n} b_i$, we have:

$$\sum_{k=1}^{10} (k^2 - 1) = \sum_{k=1}^{10} k^2 - \sum_{k=1}^{10} 1.$$

Now, use these formulas $\displaystyle\sum_{i=1}^{n} i^2 = \frac{n(n+1)(2n+1)}{6}$ and $\displaystyle\sum_{i=1}^{n} c = nc$, so, $\displaystyle\sum_{k=1}^{10} 1 = 10$, and

$$\sum_{k=1}^{10} k^2 = \frac{10(10+1)(2\times10+1)}{6} = 385. \text{ Therefore: } \sum_{k=1}^{10} (k^2 - 1) = 385 - 10 = 375.$$

# Alternate Series

- The general form of an alternating series is as follows:

$$\sum_{i=1}^{\infty}(-1)^k a_k.$$

Where $a_n \geq 0$ and the first index is arbitrary. It means that the starting term for an alternating series can have any sign.

- An alternating series $\{a_k\}_{k=1}^{\infty}$ is called convergent if:

   o $0 \leq a_{k+1} \leq a_k$, for all $k \geq 1$.

   o $a_k \to 0$, as $k \to +\infty$.

## Examples:

**Example 1.** Determine whether the following series converge or diverge:

$$\sum_{i=1}^{\infty}(-1)^i \frac{2}{i+5}$$

***Solution:*** Let $a_i = \frac{2}{i+5}$. Then:

$\frac{2}{i+5} \to 0$ as $i \to +\infty$.

In addition, $0 \leq a_{k+1} \leq a_k$:

$\frac{2}{(i+1)+5} \leq \frac{2}{i+5} \to \frac{2}{i+6} \leq \frac{2}{i+5} \to i+6 \geq i+5 \to 6 \geq 5.$

This is true. Therefore, the alternating series of the problem is convergent.

**Example 2.** Determine whether the following series converge or diverge:

$$\sum_{k=1}^{\infty} \frac{(-1)^k k}{2k+1}.$$

***Solution:*** Let $a_k = \frac{k}{2k+1}$. Then:

$\frac{k}{2k+1} \to \frac{1}{2} \neq 0$ as $i \to +\infty$.

Therefore, the alternating series of the problem is divergent.

# Chapter 13: Practices

✎ **Find the next three terms of each arithmetic sequence.**

1) $15, 11, 7, 3, -1, \ldots$

2) $-21, -14, -7, 0, \ldots$

3) $3, 6, 9, 12, 15, \ldots$

4) $4, 8, 12, 16, 20, \ldots$

✎ **Given the first term and the common difference of an arithmetic sequence find the first five terms and the explicit formula.**

5) $a_1 = 24, d = 2$

6) $a_1 = -15, d = -5$

7) $a_1 = 18, d = 10$

8) $a_1 = -38, d = -10$

✎ **Find the first five terms of the sequence.**

9) $a_1 = -120, d = -100$

10) $a_1 = 55, d = 23$

11) $a_1 = 12.5, d = 4.2$

✎ **Determine if the sequence is geometric. If it is, find the common ratio.**

12) $1, -5, 25, -125, \ldots$

13) $-2, -4, -8, -16, \ldots$

14) $4, 16, 36, 64, \ldots$

15) $-3, -15, -75, -375, \ldots$

✎ **Given the first term and the common ratio of a geometric sequence find the first five terms and the explicit formula.**

16) $a_1 = 0.8, r = -5$

17) $a_1 = 1, r = 2$

**Effortless**

**Math**

**Education**

✎ **Evaluate each geometric series described.**

18) $1, +2, +4, +8, \ldots, n = 6$    _____

19) $1, -4, +16, -64, \ldots, n = 9$    _____

20) $-2, -6, -18, -54, \ldots, n = 9$    _____

21) $2, -10, +50, -250, \ldots, n = 8$    _____

22) $1, -5, +25, -125, \ldots, n = 7$    _____

23) $-3, -6, -12, -24, \ldots, n = 9$    _____

✎ **Determine if each geometric series converges or diverges.**

24) $a_1 = -1, r = 3$

25) $a_1 = 3.2, r = 0.2$

26) $a_1 = 5, r = 2$

27) $-1, 3, -9, 27, \ldots$

28) $2, -1, \frac{1}{2}, -\frac{1}{4}, \frac{1}{8}, \ldots$

29) $81, +27, +9, +3, \ldots$

✎ **Solve.**

30) Find the 6th entry in row 8 of Pascal's triangle. _____

31) Find the 5th entry in row 9 of Pascal's triangle. _____

✎ **Solve.**

32) Write the 5th term of the expansion of $(1 - 4b^2)^4$. _____

**Effortless Math Education**

33)    Write the expansion $(2x^2 - 5)^3$. _____

## ✍ Write the terms of the series.

34) $\sum_{j=1}^{6} 4(j-2)^2$.

36) $\sum_{k=3}^{10} 2(k+3)$.

35) $\sum_{b=1}^{5} (b^2-4)^2$.

37) $\sum_{x=1}^{8} (3x^2+2)$.

## ✍ Determine whether the following series converge or diverge.

38) $\sum_{n=1}^{\infty} \frac{(-1)^n}{n^2}$

40) $\sum_{n=1}^{\infty} \frac{n^2+1}{n^3+1}$

39) $\sum_{i=1}^{\infty} (3)^i \frac{1}{i-2}$

41) $\sum_{i=1}^{\infty} \frac{(-1)^{i+3}}{i^2+4i+2}$

Effortless
Math
Education

# Chapter 13: Answers

1) $-5, -9, -13$

3) 18, 21, 24

2) 7, 14, 21

4) 24, 28, 32

5) First Five Terms: 24, 26, 28, 30, 32, Explicit: $a_n = 2n + 22$

6) First Five Terms: $-15, -20, -25, -30, -35$, Explicit: $a_n = -5n - 10$

7) First Five Terms: 18, 28, 38, 48, 58, Explicit: $a_n = 10n + 8$

8) First Five Terms: $-38, -48, -58, -68, -78$, Explicit: $a_n = -10n - 28$

9) $-120, -220, -320, -420, -520$

10) 55, 78, 101, 124, 147

11) 12.5, 16.7, 20.9, 25.1, 29.3

12) $r = -5$

13) $r = 2$

14) Not geometric

15) $r = 5$

16) First Five Terms:

    0.8, $-4$, 20, $-100$, 500

    Explicit: $a_n =$

    $0.8(-5)^{n-1}$

17) First Five Terms:

    1, 2, 4, 8, 16

    Explicit: $a_n = 2^{n-1}$

18) 63

19) 52,429

20) $-19,682$

21) $-130,208$

22) 13,021

23) $-1,533$

24) Diverges

25) Converges

26) Diverges

27) Diverges

28) Converges

29) Converges

30) 56

31) 126

32) $256b^8$

33) $8x^6 - 60x^4 + 150x^2 - 125$

34) 124

35) 619

36) 152

37) 628

38) Converge

39) Diverge

40) Diverge

41) Converge

**Effortless Math Education**

# CHAPTER

## 14 Trigonometric Functions

Math topics that you'll learn in this chapter:

☑ Trig Ratios of General Angles
☑ Trigonometric Ratios
☑ Right-Triangle Trigonometry
☑ Angles of Rotation
☑ The Unit Circle, Sine, and Cosine
☑ The Reciprocal Trigonometric Functions
☑ Function Values of Special Angles
☑ Function Values from the Calculator
☑ Reference Angles and the Calculator
☑ Coterminal Angles and Reference Angles
☑ Angles and Angle Measure
☑ Evaluating Trigonometric Function
☑ Missing Sides and Angles of a Right Triangle
☑ Arc length and Sector Area
☑ Inverse of Trigonometric Functions
☑ Solving Trigonometric Equations

187

# Trig Ratios of General Angles

- Learn common trigonometric functions:

| $\theta$ | $0°$ | $30°$ | $45°$ | $60°$ | $90°$ |
|---|---|---|---|---|---|
| $\sin\theta$ | $0$ | $\dfrac{1}{2}$ | $\dfrac{\sqrt{2}}{2}$ | $\dfrac{\sqrt{3}}{2}$ | $1$ |
| $\cos\theta$ | $1$ | $\dfrac{\sqrt{3}}{2}$ | $\dfrac{\sqrt{2}}{2}$ | $\dfrac{1}{2}$ | $0$ |
| $\tan\theta$ | $0$ | $\dfrac{\sqrt{3}}{3}$ | $1$ | $\sqrt{3}$ | Undefined |

## Examples:

**Find each trigonometric function.**

**Example 1.** What is the value of $\cos 120°$?

*Solution*: Use the following property:

$\cos(x) = \sin(90° - x)$.

Therefore:

$\cos 120° = \sin(90° - 120°) = \sin(-30°) = -\frac{1}{2}$.

**Example 2.** What is the value of $\sin(135°)$?

*Solution*: Use the following property:

$\sin(x) = \cos(90° - x)$.

Therefore:

$\sin(135°) = \cos(90° - 135°) = \cos(-45°)$.

Now use the following property:

$\cos(-x) = \cos(x)$.

Then:

$\cos(-45°) = \cos(45°) = \frac{\sqrt{2}}{2}$.

# Trigonometric Ratios

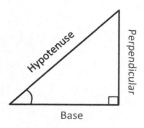

- Trigonometry is a branch of math that deals with the relationship between the angles and sides of a right-angled trigon. There are 6 trigonometric ratios: sine, cosine, tangent, secant, cosecant, and cotangent. These ratios are written as sin, cos, tan, sec, cosec or csc, and cot briefly. Trigonometric ratios could be accustomed to determine the ratios of any 2 sides out of a complete 3 sides of a right-angled trigon in terms of the respective angles.

- The 6 trigonometric ratios will be outlined as:

- sine: The ratio of the perpendicular side of the angle to the hypotenuse.
- cosine: The ratio of the side adjacent to its angle to the hypotenuse.
- tangent: The ratio of the opposite side of the angle to the adjacent side of its angle.
- cosecant: Cosecant could be defined as a multiplicative inverse of sine.
- secant: Secant could be defined as a multiplicative inverse of cosine.
- cotangent: Cotangent could be defined as the multiplicative inverse of the tangent.

## Examples:

**Example 1.** In a right-angled trigon, right-angled at $B$, the hypotenuse is 12cm, base 6cm and perpendicular is 4cm. If $\angle ACB = \theta$, then find the trigonometric ratio of $sin\,\theta$, and $cos\,\theta$.

**Solution**: We know that $sin\,\theta = \frac{perpendicular}{hypotenuse}$ and $cos\,\theta = \frac{base}{hypotenuse}$. So we put values of the hypotenuse, base, and perpendicular in these formulas to find the trigonometric ratio of $sin\,\theta$, and $cos\,\theta$: $sin\,\theta = \frac{perpendicular}{hypotenuse} \rightarrow sin\,\theta = \frac{4}{12} = \frac{1}{3} \rightarrow sin\,\theta = \frac{1}{3}$. $cos\,\theta = \frac{base}{hypotenuse} \rightarrow cos\,\theta = \frac{6}{12} = \frac{1}{2} \rightarrow cos\,\theta = \frac{1}{2}$.

**Example 2.** Find the value of $tan\,\theta$ if $sin\,\theta = \frac{10}{3}$ and $cos\,\theta = \frac{4}{3}$.

**Solution**: we know that $sin\,\theta = \frac{perpendicular}{hypotenuse}$, $cos\,\theta = \frac{base}{hypotenuse}$, and $tan\,\theta = \frac{perpendicular}{base}$. According to the question, we have a trigonometric ratio of $sin\,\theta$, and $cos\,\theta$. So, we can use the numerator to find the trigonometric ratio of $tan\,\theta$: $sin\,\theta = \frac{perpendicular}{hypotenuse} = \frac{10}{3}$, and $cos\,\theta = \frac{base}{hypotenuse} = \frac{4}{3} \rightarrow tan\,\theta = \frac{perpendicular}{base} = \frac{10}{4} = \frac{5}{2}$.

bit.ly/3Qv90fj
Find more at

# Right-Triangle Trigonometry

- Trigonometric ratios of an angle $\theta$ are the ratio of the lengths of the sides in a right triangle.

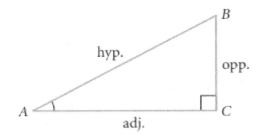

- The trigonometric functions are defined below with these abbreviations:

❖ $\sin\theta = \frac{opp.}{hyp.}$

❖ $\cos\theta = \frac{adj.}{hyp.}$

❖ $\tan\theta = \frac{opp.}{adj.}$

❖ $\cot\theta = \frac{adj.}{opp.}$

❖ $\sec\theta = \frac{hyp.}{adj.}$

❖ $\csc\theta = \frac{hyp.}{opp.}$

## Example:

For $\Delta\,ABC$, find the side length of $AB$.

*Solution:* To find $AB$, use the *cos* ratio:

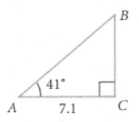

$$\cos A = \frac{adj.}{hyp.}$$

$$\cos 41^\circ = \frac{7.1}{AB}$$

$$AB = \frac{7.1}{\cos 41^\circ}$$

$$AB = \frac{7.1}{0.75}$$

$$= 9.46$$

# Angles of Rotation

- In trigonometry, an angle is defined by a ray that revolves around its endpoint. Each position of the rotated ray, relative to its starting position, creates an angle of rotation. The letter $\theta$ is used to name the angle of rotation.

- The initial position of the ray is called the initial side of the angle and the final position is called the terminal side of the angle.

- An angle is said to be in standard position when the initial side is along the positive $x$ −axis and its endpoint is at the origin.

- Angles in a standard position that have the same terminal side are coterminal.

- The reference angle, $\theta_{ref}$, is the smallest possible angle formed by the terminal side of the given angle with the $x$ −axis.

### Reference Angle Formula

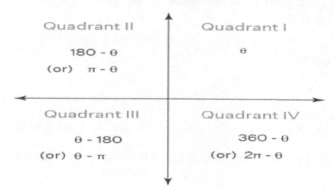

## Examples:

**Example 1.** Find the coterminal of $30°$ in the interval $-360° < \theta < 360°$.

*Solution:* You can find coterminal angles by adding or subtracting multiples of integers of $360°$.

$\theta = 30° + 360° = 390°$

$\theta = 30° - 360° = -330°$

The coterminal angle is $-330°$.

**Example 2.** Find the reference angle of $120°$.

*Solution:* You know that $120°$ lies in quadrant II. $180° - 120° = 60°$

# The Unit Circle, Sine, and Cosine

- The unit circle is a circle with a center at the origin and a radius of 1 and has the equation $x^2 + y^2 = 1$.

- If $\angle POQ$ is an angle in standard position and $P$ is the point that the terminal side of the angle intersects the unit circle and $m \angle POQ = \theta$. Then:

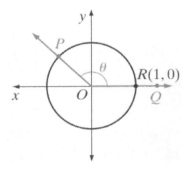

- ❖ The sine function is a set of ordered pairs $(\theta, \sin \theta)$ that $\sin \theta$ is the $y$ coordinate of $P$.
- ❖ The cosine function is the set of ordered pairs $(\theta, \cos \theta)$ that $\cos \theta$ is the $x$ −coordinate of $P$.

## Examples:

**Example 1.** If $P(\frac{\sqrt{3}}{2}, -\frac{1}{2})$ is a point on the unit circle and the terminal side of an angle in a standard position whose size is $\theta$. Find $\sin \theta$ and $\cos \theta$.

***Solution:***

$\sin \theta = y$ −coordinate of $P = -\frac{1}{2}$.

$\cos \theta = x$ −coordinate of $P = \frac{\sqrt{3}}{2}$.

**Example 2.** Does point $P\left(\frac{1}{4}, \frac{1}{4}\right)$ lie on the unit circle?

***Solution:*** The equation of a unit circle is: $x^2 + y^2 = 1$. Now substitute $x = \frac{1}{4}$ and $y = \frac{1}{4}$:

$(\frac{1}{4})^2 + (\frac{1}{4})^2 = \frac{1}{8} \neq 1$

Since, $x^2 + y^2 \neq 1$, the point $P\left(\frac{1}{4}, \frac{1}{4}\right)$ does not lie on the unit circle.

# The Reciprocal Trigonometric Functions

- The trigonometric functions that can be defined in terms of $\theta$, $\cos\theta$, and $\tan\theta$ are called the reciprocal functions.

- The secant function is the set of ordered pairs $(\theta, \sec\theta)$ for all $\theta$ for which $\cos\theta \neq 0$, $\sec\theta = \frac{1}{\cos\theta}$.

- The set of secant function values is the set of real numbers that is

  $\{x : x \geq 1 \text{ or } x \leq -1\}$.

- The cosecant function is the set of ordered pairs $(\theta, \csc\theta)$ for all $\theta$ for which $\sin\theta \neq 0$, $\csc\theta = \frac{1}{\sin\theta}$.

- The set of cosecant function values is the set of real numbers that is

  $\{x : x \geq 1 \text{ or } x \leq -1\}$.

- The cotangent function is the set of ordered pairs $(\theta, \cot\theta)$ that for all $\theta$ for which $\tan\theta$ is defined and not equal to 0, $\cot\theta = \frac{1}{\tan\theta}$, and for all $\theta$ for which $\tan\theta$ is not defined, $\cot\theta = 0$.

- The set of cotangent function values is the set of real numbers.

## Examples:

**Example 1.** Find the value of $\sec\theta$ if $\cos\theta = \frac{2}{7}$ using the reciprocal identity.

*Solution:* The reciprocal identity of $\sec$ is: $\sec\theta = \frac{1}{\cos\theta}$.

If $\cos\theta = \frac{2}{7}$, then $\sec\theta = \frac{1}{\frac{2}{7}} = \frac{7}{2}$.

**Example 2.** Simplify the function $tan(\theta)\,cot(\theta)\,sin(\theta)$.

*Solution:* The reciprocal identity of $\cot\theta$: $\cot\cot\theta = \frac{1}{tan(\theta)}$.

$$tan(\theta)\,cot(\theta)\,sin(\theta) = tan(\theta) \times \frac{1}{tan(\theta)} \times sin(\theta)$$

$$= sin(\theta)$$

# Function Values of Special Angles

- It is useful to remember the exact values of the trigonometric function summarized below:

| $\theta$ | $0°$ | $30°$ | $45°$ | $60°$ | $90°$ |
|---|---|---|---|---|---|
| $\sin\theta$ | 0 | $\dfrac{1}{2}$ | $\dfrac{\sqrt{2}}{2}$ | $\dfrac{\sqrt{3}}{2}$ | 1 |
| $\cos\theta$ | 1 | $\dfrac{\sqrt{3}}{2}$ | $\dfrac{\sqrt{2}}{2}$ | $\dfrac{1}{2}$ | 0 |
| $\tan\theta$ | 0 | $\dfrac{\sqrt{3}}{3}$ | 1 | $\sqrt{3}$ | undefined |

## Examples:

**Example 1.** Find the exact value of $\sec 45°$.

**Solution**: We know that $\sec\theta = \frac{1}{\cos\theta}$.

$$\sec 45° = \frac{1}{\cos 45°} = \frac{1}{\frac{\sqrt{2}}{2}} = \frac{2}{\sqrt{2}}$$

$$= \frac{2}{\sqrt{2}} \times \frac{\sqrt{2}}{\sqrt{2}} = \frac{2\sqrt{2}}{2} = \sqrt{2}$$

**Example 2.** Find the value of $\sin 30° \cos 60°$.

**Solution:** $\sin 30° = \frac{1}{2}$ and $\cos 60° = \frac{1}{2}$

$$\sin 30° \cos 60° = \frac{1}{2} \times \frac{1}{2} = \frac{1}{4}$$

**Example 3.** Find the value of $\cos 30° + \tan 0° + \sin 60°$.

**Solution:** $\cos 30° = \frac{\sqrt{3}}{2}, \tan 0° = 0$ and $\sin 60° = \frac{\sqrt{3}}{2}$

$$\cos 30° + \tan 0° + \sin 60° = \frac{\sqrt{3}}{2} + 0 + \frac{\sqrt{3}}{2} = \sqrt{3}$$

# Function Values from the Calculator

- You can use a calculator to find the value of trigonometric functions.

- Press MODE in your calculator. The third line of that menu is DEGREE and RADIAN. These are two common measures of angles.

- These steps follow for evaluating trigonometric functions on a scientific calculator.

   ❖ Select the correct mode (degrees or radians).
   ❖ Click the *sin*, *cos*, or *tan* button.
   ❖ Enter the angle measurement.
   ❖ Click 'enter'.

## Examples:

**Example 1.** Find $sin\ 225°$ to four decimal places.

*Solution*:

$sin\ 225° = -0.7071$

**Example 2.** Find $sec\ 44°$ to four decimal places.

*Solution*: $sec\ 44° = \frac{1}{cos\ 44°}$

$sec\ 44° = 1.3901$

**Example 3.** Find $cot\ 56°$.

*Solution*: $cot\ 56° = \frac{1}{tan\ 56°}$

$cot\ 56° = 0.67$

**Example 4.** Find $cos\ -130°$ to four decimal places.

*Solution*: $cos\ -130° = cos\ 130° = -0.6427$.

# Reference Angles and the Calculator

- If $\theta$ is the measure of an angle $90° < \theta < 360°$, the reference angle will be:

|  | Second Quadrant | Third Quadrant | Fourth Quadrant |
|---|---|---|---|
| Reference Angle | $180 - \theta$ | $\theta - 180$ | $360 - \theta$ |
| $\sin\theta$ | $\sin(180 - \theta)$ | $-\sin(\theta - 180)$ | $-\sin(360 - \theta)$ |
| $\cos\theta$ | $-\cos(180 - \theta)$ | $-\cos(\theta - 180)$ | $\cos(360 - \theta)$ |
| $\tan\theta$ | $-\tan(180 - \theta)$ | $\tan(\theta - 180)$ | $-\tan(360 - \theta)$ |

## Example:

If $\sin\theta = 0.7547$, find two positive values of $\theta$ that are less than $360°$.

**Solution:** Use the calculator to find $arc\,sin\,0.7547$.

$arc\,sin\,0.7547 = 49°$

The measure of the reference angle is $49°$. The sine is negative in the third and fourth quadrant.

In the third quadrant:
$49° = \theta - 180°$

$49° + 180° = \theta$

$\theta = 229°$

In the fourth quadrant:
$49° = 360° - \theta$

$360° - 49° = \theta$

$\theta = 311°$

# Coterminal Angles and Reference Angles

- Coterminal angles are equal angles.

- To find a coterminal of an angle, add or subtract 360 degrees (Or $2\pi$ for radians) from the given angle.

- the reference angle is the smallest angle that you can make from the terminal side of an angle with the $x$ −axis.

## Examples:

**Example 1.** Find a positive and a negative coterminal angle to angle $65°$.

*Solution*: By definition, we have:

$65° − 360° = −295°$,

$65° + 360° = 425°$.

$−295°$ and a $425°$ are coterminal with a $65°$.

**Example 2.** Find positive and negative coterminal angle to angle $\frac{\pi}{2}$.

*Solution*: According to the definition, we have:

$\frac{\pi}{2} + 2\pi = \frac{5\pi}{2}$,

$\frac{\pi}{2} − 2\pi = −\frac{3\pi}{2}$.

This means that, $\frac{5\pi}{2}$ and a $−\frac{3\pi}{2}$ are coterminal with a $\frac{\pi}{2}$.

**Example 3.** Find a positive and negative coterminal angle to angle $25°$.

*Solution*: According to the definition, we have:

$25° − 360° = −335°$,

$25° + 360° = 385°$.

$−335°$ and a $385°$ are coterminal with a $25°$.

bit.ly/3X3R2TE
Find more at

# Angles and Angle Measure

- To convert degrees to radians, use this formula: $Radian = Degrees \times \frac{\pi}{180}$.

- To convert radians to degrees, use this formula: $Degrees = Radian \times \frac{180}{\pi}$.

## Examples:

**Example 1.** Convert 120 degrees to radians.

*Solution*: Use this formula:

$Radian = Degrees \times \frac{\pi}{180}$.

Therefore:

$Radian = 120 \times \frac{\pi}{180} = \frac{120\pi}{180} = \frac{2\pi}{3}$.

**Example 2.** Convert $\frac{\pi}{3}$ to degrees.

*Solution*: Use this formula:

$Degrees = Radians \times \frac{180}{\pi}$.

Then:

$Degrees = \frac{\pi}{3} \times \frac{180}{\pi} = \frac{180\pi}{3\pi} = 60$.

**Example 3.** Convert $\frac{2\pi}{5}$ to degrees.

*Solution*: Use this formula:

$Degrees = Radians \times \frac{180}{\pi}$.

Therefore:

$Degrees = \frac{2\pi}{5} \times \frac{180}{\pi} = \frac{360\pi}{5\pi} = 72$.

**Example 4.** Convert 45 degrees to radians.

*Solution*: Use this formula:

$Radian = Degrees \times \frac{\pi}{180}$.

Therefore:

$Radian = 45 \times \frac{\pi}{180} = \frac{45\pi}{180} = \frac{\pi}{4}$.

bit.ly/3pxMlAh

Find more at

# Evaluating Trigonometric Function

- Step1: Draw the terminal side of the angle.

- Step2: Find the reference angle. (It is the smallest angle that you can make from the terminal side of an angle with the $x$ −axis.)

- Step3: Find the trigonometric function of the reference angle.

## Examples:

**Example 1.** Find the exact value of the trigonometric function. $tan\frac{4\pi}{3}$

***Solution***: Rewrite the angles for $\frac{4\pi}{3}$:

$$tan\frac{4\pi}{3} = tan\left(\frac{3\pi+\pi}{3}\right) = tan\left(\pi + \frac{1}{3}\pi\right).$$

Use the periodicity of $tan$:

$$tan(x + k \cdot \pi) = tan(x).$$

Therefore:

$$tan\left(\pi + \frac{1}{3}\pi\right) = tan\left(\frac{1}{3}\pi\right) = \sqrt{3}.$$

**Example 2.** Find the exact value of the trigonometric function. $cos\,270°$

***Solution***: Use this property:

$$cos(180° + x) = -cos(x).$$

Write $cos\,270°$ as $cos(180° + 90°)$.

Then:

$$cos(180° + 90°) = -cos(90°).$$

Recall that $cos\,90° = 0$.

Since $cos(90°) = 0$, therefore:

$$cos(270°) = -cos(90°) = 0.$$

Actually, the reference angle of $270°$ is $90°$.

# Missing Sides and Angles of a Right Triangle

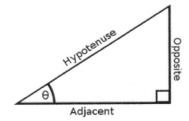

- By using Sine, Cosine or Tangent, we can find an unknown side in a right triangle when we have one length, and one angle (Apart from the right angle).

- Adjacent, Opposite and Hypotenuse, in a right triangle are shown below.

- Recall the three main trigonometric functions:

$$\text{SOH–CAH–TOA}, \quad \sin\theta = \frac{opposite}{hypotenuse}, \quad \cos\theta = \frac{adjacent}{hypotenuse}, \quad \tan\theta = \frac{opposite}{adjacent}.$$

## Examples:

**Example 1.** Find $AC$ in the following triangle. Round answers to the nearest tenth.

**Solution**: Considering that: $\sin\theta = \frac{opposite}{hypotenuse}$.

Therefore: $\sin 45° = \frac{AC}{8} \rightarrow 8 \times \sin 45° = AC$,

Now, use a calculator to find $\sin 45°$.

$\sin 45° = \frac{\sqrt{2}}{2} \cong 0.70710$.

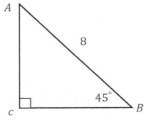

Then:

$AC = 8 \times 0.70710 = 5.6568 = 5.7$

**Example 2.** If $\tan\alpha = \frac{3}{4}$, then $\sin\alpha = ?$

**Solution**: We know that: $\tan\theta = \frac{opposite}{adjacent}$, and $\tan\alpha = \frac{3}{4}$.

Therefore, the opposite side of the angle $\alpha$ is 3 and the adjacent side is 4. Let's draw the triangle.

Using the Pythagorean theorem, we have:

$a^2 + b^2 = c^2 \rightarrow 3^2 + 4^2 = c^2 \rightarrow 9 + 16 = c^2 \rightarrow c = 5$.

Then:

$\sin\alpha = \frac{opposite}{hypotenuse} = \frac{3}{5}$.

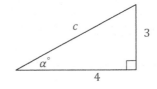

# Arc length and Sector Area

- To find a sector of a circle, use this formula:

$$\text{Area of a sector} = \pi r^2 \left(\frac{\theta}{360}\right).$$

Where $r$ is the radius of the circle and $\theta$ is the central angle of the sector.

- To find the arc of a sector of a circle, use this formula:

$$\text{Arc of a sector} = \left(\frac{\theta}{180}\right)\pi r.$$

## Examples:

**Example 1.** Find the length of the arc. Round your answers to the nearest tenth. $(\pi = 3.14)$ $r = 24cm$, $\theta = 60°$

**Solution**: Use this formula: Length of a sector $= \left(\frac{\theta}{180}\right)\pi r$.

Therefore:

Length of a sector $= \left(\frac{60}{180}\right)\pi(24) = \left(\frac{1}{3}\right)\pi(24) = 8 \times 3.14 \cong 25.1cm$.

**Example 2.** Find the area of the sector. $r = 6ft$, $\theta = 90°$, $(\pi = 3.14)$

**Solution**: Use this formula: Area of a sector $= \pi r^2 \left(\frac{\theta}{360}\right)$.

Therefore:

Area of a sector $= (3.14)(6^2)\left(\frac{90}{360}\right) = (3.14)(36)\left(\frac{1}{4}\right) = 28.26$.

**Example 3.** If the length of the arc is $18,84cm$ where $r = 4cm$. Find the area of the sector.

**Solution**: Use this formula: Arc of a sector $= \left(\frac{\theta}{180}\right)\pi r$.

Then:

$18.84 = (3.14)(4)\left(\frac{\theta}{180}\right) \rightarrow 18.84 = 12.56\left(\frac{\theta}{180}\right) \rightarrow \theta = 270°$,

Now, use this formula: Area of a sector $= \pi r^2 \left(\frac{\theta}{360}\right)$.

Therefore:

Area of a sector $= \left(\frac{270}{360}\right)(3.14)(4)^2 = 37.68cm^2$.

# Inverse of Trigonometric Functions

- The sine, cosine, and tangent functions are not one-to-one functions and do not have an inverse function. We can limit the domain of the sine, cosine, and tangent functions to form one-to-one functions that have an inverse function.

| Function with a Restricted Domain | Inverse Function |
|---|---|
| $y = \sin x$ | $y = arc\ \sin x = \sin^{-1} x$ |
| Domain $= \{x: -\frac{\pi}{2} \le x \le \frac{\pi}{2}\}$ | Domain $= \{x: -1 \le x \le 1\}$ |
| Range $= \{y: -1 \le y \le 1\}$ | Range $= \{y: -\frac{\pi}{2} \le y \le \frac{\pi}{2}\}$ |
| $y = \cos x$ | $y = arc\ \cos x$ or $y = \cos^{-1} x$ |
| Domain $= \{x: 0 \le x \le \pi\}$ | Domain $= \{x: -1 \le x \le 1\}$ |
| Range $= \{y: -1 \le y \le 1\}$ | Range $= \{y: 0 \le y \le \pi\}$ |
| $y = \tan x$ | $y = arc\ \tan x$ or $y = \tan^{-1} x$ |
| Domain $= \{x: -\frac{\pi}{2} < x < \frac{\pi}{2}\}$ | Domain $= \{x: x$ is a real number$\}$ |
| Range $= \{y: y$ is a real number$\}$ | Range $= \{y: -\frac{\pi}{2} < y < \frac{\pi}{2}\}$ |

## Examples:

**Example 1.** Determine the value of $\cos^{-1}\left(\cos \frac{13\pi}{6}\right)$.

**Solution**: $\cos^{-1}\left(\cos \frac{13\pi}{6}\right) = \cos^{-1}\left[\cos\left(2\pi + \frac{\pi}{6}\right)\right]$

$$= \cos^{-1}\left[\cos \frac{\pi}{6}\right]$$

$$= \frac{\pi}{6}$$

**Example 2.** Find the value of $\tan^{-1}(\sin 90°)$.

**Solution**: $\tan^{-1}(\sin 90°) = \tan^{-1}(1) = 45°$

# Solving Trigonometric Equations

- A trigonometric equation is an equation whose variable is expressed in terms of the value of the trigonometric function.

- A trigonometric equation can also be solved using methods for solving quadratic equations.

- A trigonometric equation may include two trigonometric functions. You can use trigonometric identities to write an equation in terms of only one of the functions.

## Examples:

**Example 1.** Find the solutions of $sin^2 \theta - 2 \sin \theta - 3 = 0$ for $0° \leq \theta < 360°$.

*Solution*:

$sin^2 \theta - 2 \sin \theta - 3 = 0$

$(sin \theta)^2 - 2(sin \theta) - 3 = 0$

$(sin \theta + 1)(sin \theta - 3) = 0$

$sin \theta = -1$ or $sin \theta = 3$

For $sin \theta = -1$, $\theta = 270°$. The equation $sin \theta = 3$ has no solution.

**Example 2.** Solve $2 sin^2(x) = 2 + cos(x)$ in the interval $[0, 2\pi)$.

*Solution:* We know $sin^2(x) = 1 - \cos^2(x)$. So:

$2(1 - \cos^2(x)) = 2 + cos(x)$

$2 - 2 \cos^2(x) = 2 + cos(x)$

$-2 \cos^2(x) - cos(x) = 0$

$2 \cos^2(x) + cos(x) = 0$

$(cos(x))(2 cos(x) + 1) = 0$

$cos(x) = 0$

$2 cos(x) + 1 = 0 \Rightarrow cos(x) = -\frac{1}{2}$

In the interval $[0, 2\pi)$, $x = \{\frac{\pi}{2}, \frac{3\pi}{2}, \frac{2\pi}{3}, \frac{4\pi}{3}\}$.

bit.ly/3DcUqUC

Find more at

# Chapter 14: Practices

## ✍ Evaluate.

1) $\sin 120° =$ _____

2) $\sin -330° =$ _____

3) $\tan -90° =$ _____

4) $\cot 90° =$ _____

5) $\cos -90° =$ _____

6) $\sec 60° =$ _____

7) $\csc 480° =$ _____

8) $\cot -135° =$ _____

## ✍ Find the given trigonometric ratio.

9) $\tan O =$ ____

10) $\sin X =$ ____

11) $\cos X =$ ____

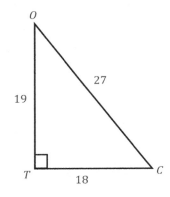

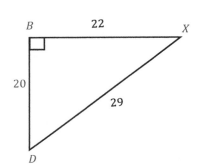

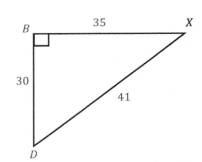

## ✍ Find the measure of each side indicated. round to the nearest tenth.

12) _____

13) _____

14) _____

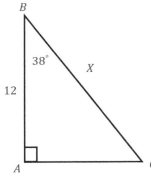

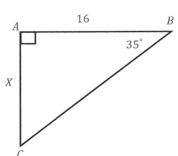

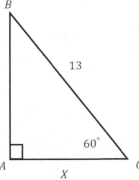

✍ **Find.**

15) Find the reference angle of $130°$. _____

16) Find the reference angle of $150°$. _____

17) Find the reference angle of $115°$. _____

✍ **Solve.**

18) If $P(-\frac{\sqrt{3}}{2}, \frac{1}{2})$ is a point on the unit circle and the terminal side of an

angle in a standard position whose size is $\theta$. Find $\sin\theta$ and $\cos\theta$.

_____

19) If $P(\frac{\sqrt{2}}{2}, -\frac{\sqrt{2}}{2})$ is a point on the unit circle and the terminal side of an

angle in a standard position whose size is $\theta$. Find $\sin\theta$ and $\cos\theta$.

_____

✍ **Solve.**

20) Find the value of $sec\ x$ if $cos\ x = \frac{3}{5}$ using the reciprocal identity.

_____

21) Find the value of $csc\ x$ if $sin\ x = \frac{2}{3}$ using the reciprocal identity.

_____

**Effortless**
**Math**
**Education**

### ✍ Find the exact value of angles.

22) $sec\,120°$

24) $sin\,45°\,cos\,30°$

23) $csc\,60°$

25) $cos\,60° + + sin\,30°$

### ✍ Find angles to four decimal places.

26) $csc\,66°$

28) $cos\,120°$

27) $sec\,56°$

29) $tan\,44°$

### ✍ Solve.

30) Find two positive values of $55°$ that are less than $360°$. _____

31) Find two positive values of $83°$ that are less than $360°$. _____

### ✍ Find a coterminal angle between $0°$ and $360°$ for each angle provided.

32) $-310° =$

34) $-440° =$

33) $-325° =$

35) $640° =$

### ✍ Find a coterminal angle between $0$ and $2\pi$ for each given angle.

36) $\frac{14\pi}{5} =$

38) $\frac{41\pi}{18} =$

37) $-\frac{16\pi}{9} =$

39) $\frac{29\pi}{12} =$

**Effortless Math Education**

✎ **Convert each degree measure into radians.**

40)  $-150° =$ ___                43) $-60° =$ ___

41)  $420° =$ ___                44) $315° =$ ___

42)  $300° =$ ___                45) $600° =$ ___

✎ **Convert each radian measure into degrees.**

46) $-\frac{16\pi}{3} =$              49) $\frac{5\pi}{9} =$

47) $-\frac{3\pi}{5} =$               50) $-\frac{\pi}{3} =$

48) $\frac{11\pi}{6} =$               51) $\frac{13\pi}{6} =$

✎ **Find the exact value of each trigonometric function.**

52) $cot -495° =$ _____          56) $cot -210° =$ _____

53) $tan\,405° =$ _____           57) $tan\frac{7\pi}{6} =$ _____

54) $cot\,390° =$ _____           58) $tan -\frac{\pi}{6} =$ _____

55) $cos -300° =$ _____

                                   59) $cot -\frac{7\pi}{6} =$ _____

Effortless
Math
Education

**Find the measure of each angle indicated.**

60)

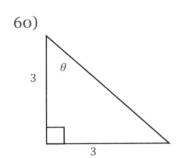

61)

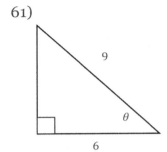

**Find the missing sides. Round answers to the nearest tenth.**

62)

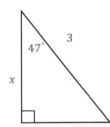

63)

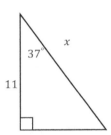

64)

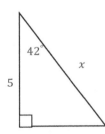

65)

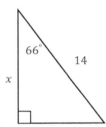

**Find the length of each arc. Round your answers to the nearest tenth.**

66) $r = 14ft,\ \theta = 45°$

67) $r = 18m,\ \theta = 60°$

68) $r = 26m,\ \theta = 90°$

69) $r = 20m,\ \theta = 120°$

✍ **Find the area of the sector. Round your answers to the nearest tenth.**

70) $r = 4m$, $\theta = 20°$

72) $r = 8m$, $\theta = 90°$

71) $r = 2m$, $\theta = 45°$

73) $r = 4m$, $\theta = 135°$

✍ **Find the value of each expression.**

74) $cos^{-1}\left(cos\frac{2\pi}{3}\right)$

76) $tan^{-1}(sec\,\pi)$

75) $tan^{-1}\left(tan\frac{3\pi}{4}\right)$

77) $sin^{-1}\left(sin\frac{11\pi}{6}\right)$

✍ **Find the value of each expression.**

78) $2\,sin^2\,x - sin\,x - 1 = 0$

79) $4\,cos^2\,x + 5\,cos\,x = 0$

80) $4\,cos^2\,x - 1 = 0$

81) $2\,sin^2\,x - 3 = 0$

Effortless
Math
Education

# Chapter 14: Answers

1) $\frac{\sqrt{3}}{2}$

2) $\frac{1}{2}$

3) Undefined

4) 0

5) 0

6) 2

7) $\frac{2\sqrt{3}}{3}$

8) 1

9) $\frac{18}{19}$

10) $\frac{20}{29}$

11) $\frac{35}{41}$

12) 15.2

13) 11.2

14) 6.5

15) $50°$

16) $30°$

17) $65°$

18) $\sin\theta = y-\text{coordinate of } P = \frac{1}{2}$

   $\cos\theta = x-\text{coordinate of } P = -\frac{\sqrt{3}}{2}$

19) $\sin\theta = y-\text{coordinate of } P = -\frac{\sqrt{2}}{2}$

   $\cos\theta = x-\text{coordinate of } P = \frac{\sqrt{2}}{2}$

20) $\frac{5}{3}$

21) $\frac{3}{2}$

22) $-2$

23) $\frac{2\sqrt{3}}{3}$

24) $\frac{\sqrt{6}}{4}$

25) $1 + \frac{\sqrt{3}}{3}$

26) 1.0946

27) 1.7882

28) $-0.5$

29) 0.9656

30) $\theta = 235°$

   $\theta = 305°$

31) $\theta = 263°$

   $\theta = 277°$

32) $50°$

33) $35°$

34) $280°$

35) $280°$

36) $\frac{4\pi}{5}$

37) $\frac{2\pi}{9}$

38) $\frac{5\pi}{18}$

39) $\frac{5\pi}{12}$

40) $-\frac{5\pi}{6}$

41) $\frac{7\pi}{3}$

42) $\frac{5\pi}{3}$

43) $-\frac{\pi}{3}$

44) $\frac{7\pi}{4}$

45) $\frac{10\pi}{3}$

46) $-960°$

47) $-108°$

48) $330°$

49) $100°$

50) $-60°$

51) $390°$

52) $1$

53) $1$

54) $\sqrt{3}$

55) $\frac{1}{2}$

56) $-\sqrt{3}$

57) $\frac{\sqrt{3}}{3}$

58) $-\frac{\sqrt{3}}{3}$

59) $-\sqrt{3}$

60) $45°$

61) $48.19°$

62) $2$

63) $13.8$

64) $6.7$

65) $5.7$

66) $11.0$

67) $18.8$

68) $40.8$

69) $41.9$

70) $2.8$

71) $1.6$

**Effortless Math Education**

72) 50.3

73) 18.8

74) $\dfrac{2\pi}{3}$

75) $-\dfrac{\pi}{4}$

76) $-\dfrac{\pi}{4}$

77) $-\dfrac{\pi}{6}$

78) $\dfrac{\pi}{2}$ or $\dfrac{7\pi}{6}$

79) $\dfrac{\pi}{2}$

80) $\dfrac{\pi}{3}$ or $\dfrac{2\pi}{3}$

81) No solution

# 15 More Topics Trigonometric Functions

Math topics that you'll learn in this chapter:

- ☑ Pythagorean Identities
- ☑ Domain and Range of Trigonometric Functions
- ☑ Cofunctions
- ☑ Law of Sines
- ☑ Law of Cosines
- ☑ Fundamental Trigonometric Identities
- ☑ Sum and Difference Identities
- ☑ Double-Angle and Half-Angle Identities

213

# Pythagorean Identities

- An identity is an equation that is true for all variable values for which the variable expressions are defined.

- Since the identity $sin^2 \theta + cos^2 \theta = 1$ is based on the Pythagorean theorem, we refer to it as the Pythagorean identity.

- Two related Pythagorean identities can be written by dividing both sides of the equation by the same expression:

  ❖ $1 + tan^2 \theta = sec^2 \theta$

  ❖ $cot^2 \theta + 1 = csc^2 \theta$

## Examples:

**Example 1.** Verify that $sin^2 \frac{\pi}{4} + cos^2 \frac{\pi}{4} = 1$.

**Solution**: $cos \frac{\pi}{4} = \frac{\sqrt{2}}{2}$ and $sin \frac{\pi}{4} = \frac{\sqrt{2}}{2}$:

$$sin^2 \frac{\pi}{4} + cos^2 \frac{\pi}{4} = \left(\frac{\sqrt{2}}{2}\right)^2 + \left(\frac{\sqrt{2}}{2}\right)^2 = \frac{1}{2} + \frac{1}{2} = 1$$

**Example 2.** If $cos x = \frac{3}{5}$ and $x$ is in the 1st quadrant, find $sin x$.

**Solution**: Use Pythagorean identity:

$$sin^2 x = 1 - cos^2 x$$

$$sin(x) = \pm\sqrt{1 - cos^2(x)} = \pm\sqrt{1 - \left(\frac{3}{5}\right)^2} = \pm\frac{4}{5}$$

Since $x$ is in the first quadrant, $sin x$ is positive. So $sin x = \frac{4}{5}$.

**Example 3.** Use a Pythagorean identity to simplify the $14 + 5\,cos^2(x) + 5\,sin^2(x)$.

**Solution**: The Pythagorean identity is $sin^2 \theta + cos^2 \theta = 1$.

$$14 + 5\,cos^2(x) + 5\,sin^2(x) = 14 + 5(cos^2(x) + sin^2(x))$$

$$= 14 + 5(1) = 19.$$

# Domain and Range of Trigonometric Functions

- The domain of the sine function and cosine function is the set of real numbers.

- The range of the sine function and cosine function is the set of real numbers $[-1, 1]$.

- The domain of the tangent function is the set of real numbers except for $\frac{\pi}{2} + n\pi$ for all integral values of $n$.

- The range of the tangent function is the set of all real numbers.

- The domain of the cotangent function is the set of real numbers except for $n\pi$ for all integral values of $n$.

- The range of the cotangent function is the set of all real numbers.

- The domain of the secant function is the set of real numbers except for $\frac{\pi}{2} + n\pi$ for all integral values of $n$.

- The range of the secant function is the set of real numbers $(-\infty, -1] \cup [1, \infty)$.

- The domain of the cosecant function is the set of real numbers except for $n\pi$ for all integral values of $n$.

- The range of the cosecant function is the set of real numbers $(-\infty, -1] \cup [1, \infty)$.

## Examples:

**Example 1.** Find the range of $y = 4 \tan x$.

*Solution*: The range of $y = 4 \tan x$ is $(-\infty, +\infty)$.

**Example 2.** Find the domain and range of $y = \sin x - 4$

*Solution*: The range of $\sin x$ is $[-1, 1]$.

$-1 \le \sin x \le 1 \Rightarrow -1 - 4 \le \sin x - 4 \le 1 - 4 \Rightarrow -5 \le y \le -3$

The domain is $(-\infty, +\infty)$.

**Example 3.** Find the domain of $y = 3 \cos x + 4$.

*Solution*: The domain of $y = 3 \cos x + 4$ is $(-\infty, +\infty)$.

bit.ly/3ZzdrKi

Find more at

# Cofunctions

- Cofunction identities show the relationship between complementary angles and trigonometric functions.

- The following equations are cofunction identities:

  ❖ $sin\,\theta = cos(90° - \theta)$
  ❖ $cos\,\theta = sin(90° - \theta)$
  ❖ $cot\,\theta = tan(90° - \theta)$
  ❖ $tan\,\theta = cot(90° - \theta)$
  ❖ $csc\,\theta = sec(90° - \theta)$
  ❖ $sec\,\theta = csc(90° - \theta)$

## Examples:

**Example 1.** Find the value of acute angle $x$, if $sin\,x = cos\,30°$.

*Solution*: Use cofunction identity, $sin\,\theta = cos(90° - \theta)$.

$sin\,x = cos\,30°$

$cos(90° - x) = cos\,30°$

$90° - x = 30°$

$x = 60°$

**Example 2.** Find the value of $x$, if $sec(8x) = csc(x + 18°)$, where $8x$ is an acute angle.

*Solution*: Use cofunction identity, $sec\,\theta = csc(90° - \theta)$.

$sec(8x) = csc(x + 18°)$

$csc(90° - 8x) = csc(x + 18°)$

$90° - 8x = x + 18°$

$90° - 18° = x + 8x$

$72° = 9x$

$x = 8°$

**Example 3.** Find the value of $sin\,135°$ using cofunction identities.

*Solution*: Use cofunction identity, $sin\,\theta = cos(90° - \theta)$.

$$sin\,135° = cos(90° - 135°)$$
$$= cos(-45°) = cos\,45° = \frac{\sqrt{2}}{2}.$$

# Law of Sines

- When you know the measures of 2 factors in a triangle as well as one of the sides, as in the case of ASA or AAS, you can use the law of sines to determine the measures of the other two angles and the other side.

- For $\Delta\ ABC$, the Law of Sines states the following:

❖ $\dfrac{a}{\sin A} = \dfrac{b}{\sin B} = \dfrac{c}{\sin C}$

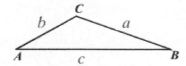

## Examples:

**Example 1.** For a triangle, it is given $a = 12cm$, $c = 14.5cm$ and angle $C = 54°$. Find the angle $A$ of the triangle.

*Solution*: Use the Law of Sines to find the side of $a$.

$\dfrac{a}{\sin A} = \dfrac{c}{\sin C}$

$\dfrac{12}{\sin A} = \dfrac{14.5}{\sin 54°} \rightarrow \sin A = \dfrac{12 \times \sin 54°}{14.5}$

$\sin A = \dfrac{12 \times 0.8}{14.5} = \dfrac{9.6}{14.5} = 0.66 \rightarrow A = 42.3°$

**Example 2.** In the $ABC$ triangle, find side $a$.

*Solution:* Use the Law of Sines to find side $a$.

$\dfrac{a}{\sin 30°} = \dfrac{54}{\sin 20°} \rightarrow a = \dfrac{54 \times \sin 30°}{\sin 20°}$

$a = \dfrac{54 \times 0.5}{0.34} = \dfrac{27}{0.34} = 79.41$

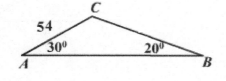

**Example 3.** For a triangle, it is given $A = 62°$, $B = 55°$ and $c = 5cm$. Find side $b$ of the triangle.

*Solution*: Use the Law of Sines to find side $b$.
$62° + 55° + C = 180° \rightarrow C = 63°$

$\dfrac{b}{\sin B} = \dfrac{c}{\sin C} \rightarrow \dfrac{b}{\sin 55°} = \dfrac{5}{\sin 63°} \rightarrow b = \dfrac{\sin 55° \times 5}{\sin 63°} \rightarrow b = 4.59cm$

bit.ly/3ZEU4zq
Find more at

# Law of Cosines

- The Law of Cosines can be used to find the size of the third side of a triangle when the size of the two sides and the included angle are known.

- The law of cosines can also be used to find the measure of each angle of a triangle when the measures of the three sides are known.

- The Law of Cosines states the following:

❖ $a^2 = b^2 + c^2 - 2bc \cos A \rightarrow \cos A = \frac{b^2+c^2-a^2}{2bc}$

❖ $b^2 = a^2 + c^2 - 2ac \cos B \rightarrow \cos B = \frac{a^2+c^2-b^2}{2ac}$

❖ $c^2 = a^2 + b^2 - 2ab \cos C \rightarrow \cos C = \frac{a^2+b^2-c^2}{2ab}$

## Examples:

**Example 1.** Find angle $B$ in the $ABC$ triangle.

*Solution*: Use the Law of Cosines to find angle $B$.

$$\cos B = \frac{a^2 + c^2 - b^2}{2ac}$$

$$\cos B = \frac{8^2 + 10^2 - 14^2}{2 \times 8 \times 10}$$

$$= \frac{64 + 100 - 196}{160} = -0.2$$

$B = 101.54°$

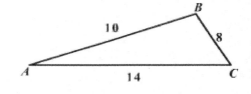

**Example 2.** For $\Delta ABC$, it is given $a = 9$, $b = 7$ and $c = 6cm$. Find angle $C$ of the triangle.

*Solution*: Use the Law of Cosines to find angle $C$.

$$\cos C = \frac{a^2+b^2-c^2}{2ab} \rightarrow \cos C = \frac{9^2+7^2-6^2}{2\times9\times7} = \frac{81+49-36}{126} \cong 0.746$$

$$C = 41.75°$$

# Fundamental Trigonometric Identities

- The following equations are important trigonometric identities:

  ❖ $tan\,\theta = \frac{sin\,\theta}{cos\,\theta}$

  ❖ $cot\,\theta = \frac{cos\,\theta}{sin\,\theta}$

  ❖ $csc\,\theta = \frac{1}{sin\,\theta}$

  ❖ $sec\,\theta = \frac{1}{cos\,\theta}$

  ❖ $cot\,\theta = \frac{1}{tan\,\theta}$

  ❖ $cos^2\,\theta + sin^2\,\theta = 1$

  ❖ $sin^2\,\theta = 1 - cos^2\,\theta$

  ❖ $cos^2\,\theta = 1 - sin^2\,\theta$

  ❖ $tan^2\,\theta + 1 = sec^2\,\theta$

  ❖ $1 + cot^2\,\theta = csc^2\,\theta$

- You can use fundamental identities to rewrite trigonometric expressions in terms of a single trigonometric function.

## Examples:

**Example 1.** Confirm the identity $cos\,\theta + sin\,\theta\,tan\,\theta = sec\,\theta$.

**Solution**: You can use fundamental trigonometric identities to solve this problem:

$$(cos\,\theta) + (sin\,\theta)\left(\frac{sin\,\theta}{cos\,\theta}\right) = sec\,\theta$$

$$\frac{cos^2\,\theta + sin^2\,\theta}{cos\,\theta} = sec\,\theta$$

$$\frac{1}{cos\,\theta} = sec\,\theta$$

$$sec\,\theta = sec\,\theta.$$

**Example 2.** Find the value of $tan\,\theta$ using $cot\,\theta = \frac{3}{5}$.

**Solution**: Use fundamental trigonometric identities to solve this problem:

$$cot\,\theta = \frac{1}{tan\,\theta} = \frac{1}{\frac{3}{5}} = \frac{5}{3}.$$

bit.ly/3QNo3RI

Find more at

# Sum and Difference Identities

- The sum and difference formulas in trigonometry are used to find the value of trigonometric functions at certain angles.

- These formulas help us evaluate the value of trigonometric functions at angles that can be expressed as the sum or difference of certain angles $0°$, $30°$, $45°$, $60°$, $90°$, and $180°$.

- The sum and difference formulas are as follows:

  ❖ $sin(A + B) = sin A \cos B + \cos A \sin B$
  ❖ $sin(A - B) = sin A \cos B - \cos A \sin B$
  ❖ $cos(A + B) = \cos A \cos B - sin A \sin B$
  ❖ $cos(A - B) = \cos A \cos B + sin A \sin B$
  ❖ $tan(A + B) = \frac{tan(A) + tan(B)}{1 - tan(A)\,tan(B)}$
  ❖ $tan(A - B) = \frac{tan(A) - tan(B)}{1 - tan(A)\,tan(B)}$

## Examples:

**Example 1.** Find the value of $sin(120° + 45°)$.

*Solution*: Use the sum and difference formula:

$$sin(120° + 45°) = sin(120°)\cos(45°) + \cos(120°)\sin(45°)$$

$$= \left(\tfrac{\sqrt{3}}{2}\right)\left(\tfrac{\sqrt{2}}{2}\right) + \left(-\tfrac{1}{2}\right)\left(\tfrac{\sqrt{2}}{2}\right)$$

$$= \tfrac{\sqrt{6}}{4} - \tfrac{\sqrt{2}}{4} = \tfrac{\sqrt{6} - \sqrt{2}}{4}.$$

**Example 2.** Fin d the value of $cos\,105°$.

*Solution:* Use the sum and difference formula:

$$cos\,105° = cos(60° + 45°) = \cos 60° \cos 45° - sin 60° \sin 45°$$

$$= \left(\tfrac{1}{2}\right)\left(\tfrac{\sqrt{2}}{2}\right) - \left(\tfrac{\sqrt{3}}{2}\right)\left(\tfrac{\sqrt{2}}{2}\right) = \tfrac{\sqrt{2}}{4} - \tfrac{\sqrt{6}}{4} = \tfrac{\sqrt{2} - \sqrt{6}}{4}.$$

# Double-Angle and Half-Angle Identities

- Double-angle formulas are used for trigonometric ratios of double angles in terms of trigonometric ratios of single angles.

- The double-angle formulas are as follows:

  ❖ $sin\,2\theta = 2\,sin\,\theta\,cos\,\theta$

  ❖ $sin\,2\theta = \frac{2\,tan\,\theta}{1+tan^2\,\theta}$

  ❖ $cos\,2\theta = cos^2\,\theta - sin^2\,\theta$

  ❖ $cos\,2\theta = 1 - 2\,sin^2\,\theta$

  ❖ $cos\,2\theta = 2\,cos^2\,\theta - 1$

  ❖ $cos\,2\theta = \frac{1-tan^2\,\theta}{1+tan^2\,\theta}$

  ❖ $tan\,2\theta = \frac{2\,tan\,\theta}{1-tan^2\,\theta}$

- The half-angle formulas are as follows:

  ❖ $sin\frac{\theta}{2} = \pm\sqrt{\frac{1-cos\,\theta}{2}}$

  ❖ $cos\frac{\theta}{2} = \pm\sqrt{\frac{1+cos\,\theta}{2}}$

  ❖ $tan\frac{\theta}{2} = \pm\sqrt{\frac{1-cos\,\theta}{1+cos\,\theta}} = \frac{sin\,\theta}{1+cos\,\theta} = \frac{1-cos\,\theta}{sin\,\theta}$

## Examples:

**Example 1.** If $tan\,\theta = \frac{4}{3}$, find the values of $cos\,2\theta$.

*Solution*: Use the double-angle formulas:

$$cos\,2\theta = \frac{1-tan^2\,\theta}{1+tan^2\,\theta} = \frac{1-\left(\frac{4}{3}\right)^2}{1+\left(\frac{4}{3}\right)^2} = \frac{1-\frac{16}{9}}{1+\frac{16}{9}} = \frac{-\frac{7}{9}}{\frac{25}{9}} = -\frac{7}{25}$$

**Example 2.** If $cos\,\theta = \frac{1}{2}$, find the values of $sin\frac{\theta}{2}$.

*Solution*: Use the half-angle formulas:

$$sin\frac{\theta}{2} = \pm\sqrt{\frac{1-cos\,\theta}{2}} = \pm\sqrt{\frac{1-\frac{1}{2}}{2}} = \pm\sqrt{\frac{\frac{1}{2}}{2}} = \pm\sqrt{\frac{1}{4}} = \pm\frac{1}{2}$$

# Chapter 15: Practices

✎ **Simplify each trigonometric expression using Pythagorean identities.**

1) $(sin\ x + cos\ x)^2 = $ _____

2) $(1 + cot^2\ x) sin^2\ x = $ _____

3) $csc^2\ x - cot^2\ x = $ _____

4) $2\ sin^2\ x + cos^2\ x = $ _____

✎ **Find the domain and range of functions.**

5) $y = cos\ x - 4$

Domain: _____

Range: _____

6) $y = sin\ x - 3$

Domain: _____

Range: _____

7) $y = \dfrac{1}{2 - sin\ 2x}$

Domain: _____

Range: _____

✎ **Solve using cofunction identities.**

8) $sin\ x = cos\ 35°$

9) $cot\ x = tan\ 80°$

10) $csc\ 68° = sec\ x$

11) $sec(3x) = csc(x + 22°)$

✎ **Find each measurement indicated. Round your answers to the nearest tenth.**

12) _____

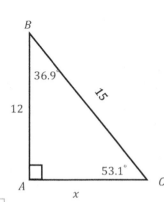

13) _____

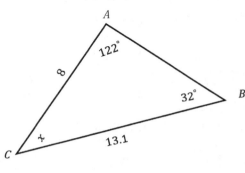

14) $m\angle C = 14°$, $m\angle A = 24°$, $c = 9$ _____

15) $m\angle C = 125°$, $b = 8$, $c = 24$ _____

✎ **Find each measurement indicated. Round your answers to the nearest tenth.**

16) _____

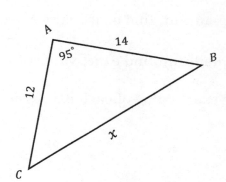

17) _____

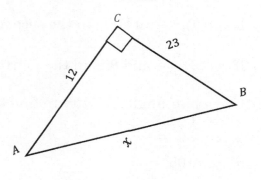

18) In $\triangle ABC$, $a = 15cm$, $b = 10cm$, $c = 7cm$ _____

19) In $\triangle ABC$, $a = 18cm$, $b = 15cm$, $c = 11cm$ _____

✎ **Simplify each expression using trigonometric fundamental identities.**

20) $cot\, x\, sec\, x\, sin\, x =$ _____

21) $sin\, x\, cos^2\, x - sin\, x =$ _____

22) $tan^3\, x\, csc^3\, x =$ _____

23) $\frac{tan\, x}{sec\, x} =$ _____

Effortless
Math
Education

✎ **Find the value of angles.**

24) $cos\ 75°$

25) $sin\ 30°$

26) $sin\ (30° + 45°)$

27) $cos(-15°)$

28) $tan\ (75°)$

29) $sin(-75°)$

✎ **Solve.**

30) If $sin(\theta) = \frac{2}{5}$ and $\theta$ is in the second quadrant, find exact values for $cos(2\theta)$.

31) If $cos(\theta) = \frac{4}{5}$ and $\theta$ is in the fourth quadrant, find exact values for $sin(2\theta)$.

32) Use a half-angle identity to find the exact value of each expression.

    a) $cos\ 30° =$

    b) $cos\ 105° =$

✎ **Calculate the angles labelled $\theta$ in each triangle.**

33) _____                    34) _____

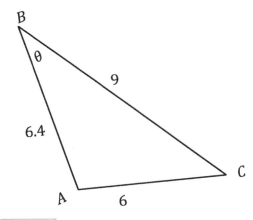

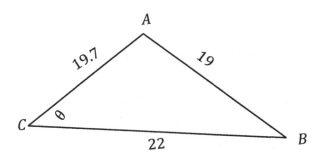

# Chapter 15: Answers

1) $1 + 2 \sin x \cos x$

2) $1$

3) $1$

4) $1 + \sin^2 x$

5) Domain: $x \in R$

   Range: $[-5, -3]$

6) Domain: $x \in R$

   Range: $[-4, -2]$

7) Domain: $(-\infty, +\infty)$

   Range: $[\frac{1}{3}, 1]$

8) $55°$

9) $10°$

10) $22°$

11) $17°$

12) $x = 9$

13) $x = 26°$

14) $m\angle B = 142°, a = 15.1, b = 22.9$

15) $m\angle A = 39.2°, m\angle B = 15.8°, a = 18.6$

16) $x = 19.2$

17) $x = 26$

18) $m\angle A = 122.9°, m\angle B = 34.1°, m\angle C = 23°$

19) $m\angle A = 86.1°, m\angle B = 56.3°, m\angle C = 37.6°$

20) $1$

21) $- \sin^3 x$

22) $\sec^3 x$

23) $\sin x$

24) $\frac{\sqrt{6} - \sqrt{2}}{4}$

25) $\frac{1}{2}$

26) $\frac{\sqrt{6} + \sqrt{2}}{4}$

27) $\frac{\sqrt{6} + \sqrt{2}}{4}$

28) $2 + \sqrt{3}$

29) $\frac{-\sqrt{6} - \sqrt{2}}{4}$

30) $\frac{17}{25}$

31) $-\frac{24}{25}$

32) $a = \frac{\sqrt{3}}{2}, b = \frac{\sqrt{2} - \sqrt{6}}{4}$

33) $\theta = 41.74°$

34) $\theta \cong 53.87°$

**Effortless Math Education**

# 16 Graphs of Trigonometric Functions

Math topics that you'll learn in this chapter:

☑ Graph of the Sine Function
☑ Graph of the Cosine Function
☑ Amplitude, Period, and Phase Shift
☑ Writing the Equation of a Sine Graph
☑ Writing the Equation of a Cosine Graph
☑ Graph of the Tangent Function
☑ Graph of the Cosecant Function
☑ Graph of the Secant Function
☑ Graph of the Cotangent Function
☑ Graph of Inverse of the Sine Function
☑ Graph of Inverse of the Cosine Function
☑ Graph of Inverse of the Tangent Function
☑ Sketching Trigonometric Graphs

# Graph of the Sine Function

-   The sine function is a set of ordered pairs of real numbers. Each ordered pair can be shown as a point on the coordinate plane.

-   To graph the sine function, we plot a portion of the graph using a subset of the real numbers in the interval $0 \leq x \leq 2\pi$.

-   We can see how $x$ and $y$ change by using the graph:

    ❖  By increasing $x$ from $0$ to $\frac{\pi}{2}$, $y$ increases from $0$ to $1$.

    ❖  By increasing $x$ from $\frac{\pi}{2}$ to $\pi$, $y$ decreases from $1$ to $0$.

    ❖  By increasing $x$ from $\pi$ to $\frac{3\pi}{2}$, $y$ continues to decrease from $0$ to $-1$.

    ❖  By increasing $x$ from $\frac{3\pi}{2}$ to $2\pi$, $y$ increases from $-1$ to $0$.

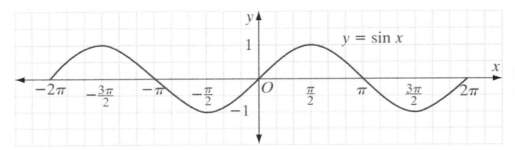

## Example:

In the interval $-2\pi \leq x \leq 0$, for what values of $x$ do $y = \sin x$ increase?

**Solution**: From the graph, we can see $y = \sin x$ increases in the interval $-2\pi \leq x \leq -\frac{3\pi}{2}$ and $-\frac{\pi}{2} \leq x \leq 0$.

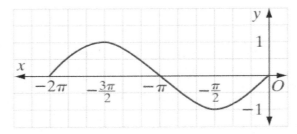

*EffortlessMath.com*

# Graph of the Cosine Function

- The cosine function is a set of ordered pairs of real numbers.

- To graph the cosine function, we plot a portion of the graph using a subset of the real numbers in the interval $0 \leq x \leq 2\pi$.

- From the graph, we can know how $x$ and $y$ change:

  ❖ By increasing $x$ from 0 to $\frac{\pi}{2}$, $y$ decreases from 1 to 0.

  ❖ By increasing $x$ from $\frac{\pi}{2}$ to $\pi$, $y$ decreases from 0 to $-1$.

  ❖ By increasing $x$ from $\pi$ to $\frac{3\pi}{2}$, $y$ increases from $-1$ to 0.

  ❖ By increasing $x$ from $\frac{3\pi}{2}$ to $2\pi$, $y$ increases from 0 to 1.

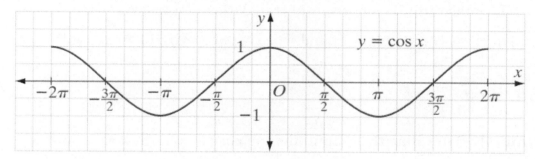

## Example:

For what values of $x$ in the interval $0 \leq x \leq 2\pi$, does $y = \cos x$ have a maximum value?

**Solution:** From the graph, we can see $y = \cos x$ at $x = 0$, and $x = 2\pi$ has a maximum value of 1.

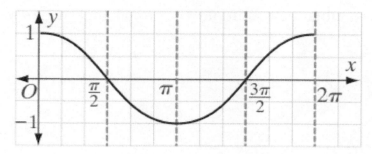

# Amplitude, Period, and Phase Shift

- For the function $y = a \sin x$, the maximum value of the function is $|a|$ and the minimum value of the function is $-|a|$.

- For the function $y = a \cos x$, the maximum value of the function is $|a|$ and the minimum value of the function is $-|a|$.

- The amplitude of a periodic function is the absolute value of half the difference between the maximum and minimum value of $y$.

- The difference between the $x$ −coordinates of the endpoints of the interval for one graph cycle is the graph period.

- The period of $y = \cos bx$ and $y = \sin bx$ is $|\frac{2\pi}{b}|$.

- The phase shift is a horizontal translation of a trigonometric function.

- For the $y = a \sin b(x + c)$ and $y = a \cos b(x + c)$, the phase shift is $-c$.

## Examples:

**Example 1.** Determine the amplitude, the period, and the phase shift of

$y = -9 \cos(8x + \pi) - 8$.

*Solution*: Amplitude= 9

Period $= \frac{\pi}{4}$

Phase shift $= -\frac{\pi}{8}$

**Example 2.** Determine the amplitude, the period, and the phase shift of

$y = \sin(3x - 4) + 5$.

*Solution*: Amplitude $= 1$

Period $= \frac{2\pi}{3}$

Phase shift $= \frac{4}{3}$

# Writing the Equation of a Sine Graph

- Using these steps, we can write an equation for the sine graph:

  ❖ Find $a$ by identifying the maximum and minimum values of $y$ for the function. $a = \frac{maximum - minimum}{2}$

  ❖ Define a basic cycle of the sine graph that starts at $y = 0$, increases to a maximum value, decreases to 0, continues to decrease to the minimum value, and then increases to 0. Find the $x$ −coordinates of the endpoints of this cycle. Write the domain of a cycle in interval notation, $x_0 \le x \le x_1$ or $[x_0, x_1]$.

  ❖ The period of one cycle is $\frac{2\pi}{b} = x_1 - x_0$. Use this formula to find $b$.

  ❖ The $c$ value is the opposite of the lower endpoint of the interval of the basic cycle: $c = -x_0$.

## Example:

Determine the equation of the graph below in the form $y = a \sin bx$.

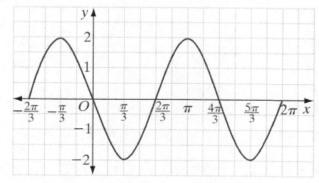

*Solution*: The max and min values of $y$ are 2 and $-2$. So, $a = \frac{2-(-2)}{2} = 2$.

The cycle of the curve is in the interval $-\frac{2\pi}{3} \le x \le \frac{2\pi}{3}$. The period is $\frac{2\pi}{3} - (-\frac{2\pi}{3})$ or $\frac{4\pi}{3}$. So $\frac{2\pi}{b} = \frac{4\pi}{3} \rightarrow b = \frac{3}{2}$.

The $c$ value is $-(-\frac{2\pi}{3})$, therefore the phase shift is $\frac{2\pi}{3}$.

The equation of the graph is $y = 2 \sin \frac{3}{2}\left(x + \frac{2\pi}{3}\right)$.

# Writing the Equation of a Cosine Graph

- Using these steps, we can write an equation for the cosine graph:

  ❖ Find $a$ by identifying the maximum and minimum values of $y$ for the function. $a = \frac{maximum - minimum}{2}$.

  ❖ Define one basic cycle of the graph that starts at the maximum value, decreases to 0, continues to decrease to the minimum value, increases to 0, and then increases to the maximum value. Find the $x$ −coordinates of the endpoints of this cycle. Write the domain of a cycle in interval notation, $x_0 \le x \le x_1$ or $[x_0, x_1]$.

  ❖ The period of one cycle is $\frac{2\pi}{b} = x_1 - x_0$. Use this formula to find $b$.

  ❖ The $c$ value is the opposite of the lower endpoint of the interval of the basic cycle: $c = -x_0$.

## Example:

Determine the equation of the graph below in the form $y = a\,\cos bx$.

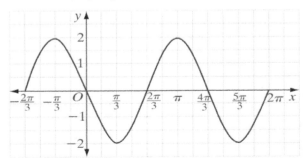

*Solution*: The max and min values of $y$ are 2 and $-2$. So, $a = \frac{2-(-2)}{2} = 2$.

The cycle of the curve is in the interval $-\frac{\pi}{3} \le x \le \pi$. The period is $\pi - (-\frac{\pi}{3})$ or $\frac{4\pi}{3}$.

So, $\frac{2\pi}{b} = \frac{4\pi}{3} \rightarrow b = \frac{3}{2}$.

The $c$ value is $-(-\frac{\pi}{3})$, therefore the phase shift is $\frac{\pi}{3}$.

The equation of the graph is $y = 2\cos\frac{3}{2}\left(x + \frac{\pi}{3}\right)$.

# Graph of the Tangent Function

- The tangent graph is a curve that increases through negative values of $\tan x$ to 0 and then continues to increase through positive values.
- The graph is discontinuous at odd multiples of $\frac{\pi}{2}$ and then repeats the same pattern.
- The $tan$ graph displays a vertical line at $x = \frac{\pi}{2}$ and at every value of $x$ that is an odd multiple of $\frac{\pi}{2}$. These lines are vertical asymptotes.
- The one complete cycle of the curve is in the interval from $x = -\frac{\pi}{2}$ to $x = \frac{\pi}{2}$ and the period of the curve is $\pi$.

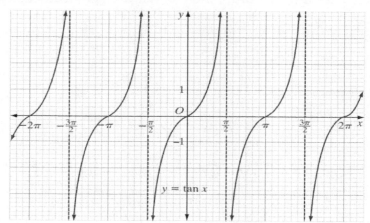

## Example:

Draw the graph of $y = \tan\left(x - \frac{\pi}{4}\right)$ in the interval of $-\frac{\pi}{4} < x < \frac{3\pi}{4}$.

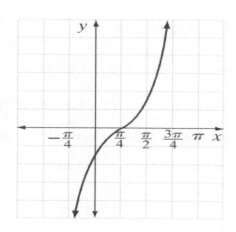

**Solution**: The graph of $y = \tan\left(x - \frac{\pi}{4}\right)$ is the graph of $y = \tan x$ but the phase shift is $\frac{\pi}{4}$.

# Graph of the Cosecant Function

- The cosecant function is identified in terms of the sine function: $csc\ x = \frac{1}{sin\ x}$.

- To draw the cosecant function graph, use the reciprocals of the sine function values.

- There are reciprocals of the sine function for $-1 \leq sin\ x < 0$, and for $0 < sin\ x < 1$. So, $-\infty < csc\ x \leq -1$, $1 \leq csc\ x < \infty$.

- For values of $x$ that are multiples of $\pi$, $sin\ x = 0$, and $csc\ x$ is not defined.

- For integral values of $n$, the vertical lines on the graph are asymptote at $x = n\pi$.

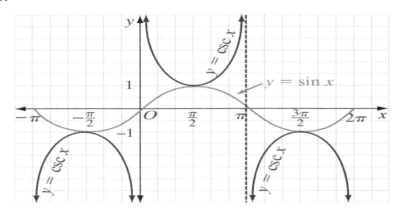

## Example:

Draw one period of $y = -3\ csc(4x)$.

**Solution**: Draw a graph of the function $y = -3\ sin(4x)$. Sketch vertical asymptotes and fill in the cosecant curve in between the asymptotes.

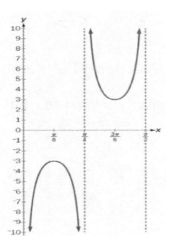

# Graph of the Secant Function

- The secant function is identified in terms of the cosine function: $sec\,x = \frac{1}{cos\,x}$

- To draw the secant function, use reciprocals of the cosine function values.

- There are reciprocals of the cosine function for $-1 \le cos\,x < 0$, and for $0 < cos\,x \le 1$. So, $-\infty < sec\,x \le -1$, $1 \le sec\,x < \infty$.

- For $x$ values that are odd multiples of $\frac{\pi}{2}$, $cos\,x = 0$, and $sec\,x$ is not defined.

- For integral values of $n$, the vertical lines on the graph are asymptote at $x = \frac{\pi}{2} + n\pi$.

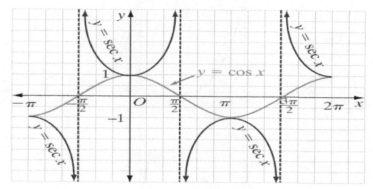

## Example:

Draw one period of $y = sec\left(2x - \frac{\pi}{2}\right) + 3$.

***Solution***: Draw a graph of the function $y = cos\left(2x - \frac{\pi}{2}\right) + 3$. Sketch vertical asymptotes and fill in the secant curve in between the asymptotes.

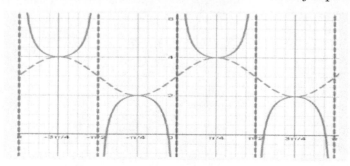

bit.ly/3ZFAQK2

Find more at

# Graph of the Cotangent Function

- The cotangent function is identified in terms of the tangent function: $cot\ x = \frac{1}{tan\ x}$.

- To draw the cotangent function, use the reciprocals of the tangent function values.

- For values of $x$ that are multiples of $\pi$, $tan\ x = 0$, and $cot\ x$ is not defined.

- For values of $x$ that $tan\ x$ is not defined, $cot\ x = 0$.

- For integral values of $n$, the vertical lines on the graph are asymptote at $x = n\pi$.

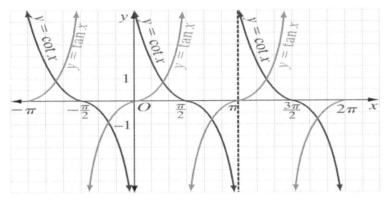

## Example:

Draw one period of $y = cot\left(x + \frac{\pi}{2}\right)$.

**Solution**: Draw a graph of the function $y = tan\left(x + \frac{\pi}{2}\right)$. Sketch vertical asymptotes and fill in the cotangent curve in between the asymptotes.

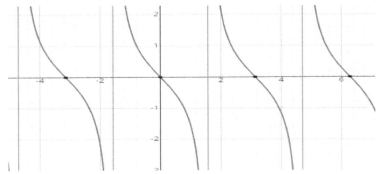

# Graph of Inverse of the Sine Function

- If we limit the domain of the sine function to $-\frac{\pi}{2} \le x \le \frac{\pi}{2}$, that subset of the sine function is a one-to-one function and has an inverse function.

- When we reflect that subset on the line $y = x$, the image of the function is $y = arc\,sin\,x$ or $y = sin^{-1}\,x$.

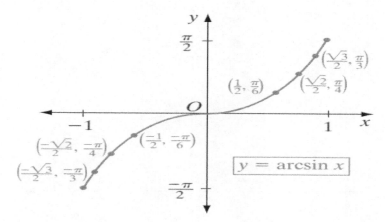

## Example:

Draw the function $y = 3\,sin^{-1}(x+1)$.

**Solution**: The graph of $y = 3\,sin^{-1}(x+1)$ is the graph of $y = sin^{-1}\,x$ but the phase shift is $-1$.

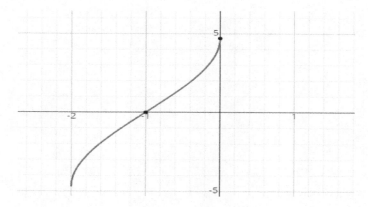

# Graph of Inverse of the Cosine Function

- If we limit the domain of the cosine function to $0 \leq x \leq \pi$, that subset of the cosine function is a one-to-one function and has an inverse function.

- When we reflect that subset on the line $y = x$, the image of the function is $y = arc \, cos \, x$ or $y = cos^{-1} x$.

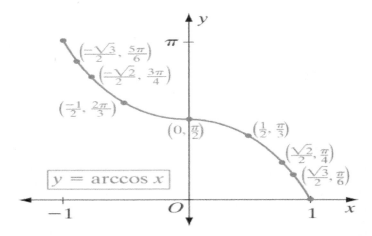

## Example:

Draw the function $y = 2 \, cos^{-1}(x - 1)$.

**Solution**: The graph of $y = 2 \, cos^{-1}(x - 1)$ is the graph of $y = cos^{-1} x$ but the phase shift is 1.

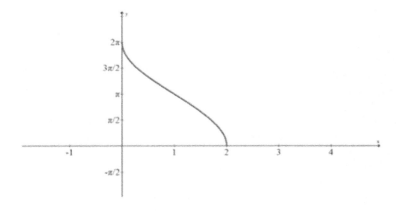

# Graph of Inverse of the Tangent Function

- If we limit the domain of the tangent function to $-\frac{\pi}{2} < x < \frac{\pi}{2}$, that subset of the tangent function is a one-to-one function and has an inverse function.

- When we reflect that subset on the line $y = x$, the image of the function is $y = arc\,tan\,x$ or $y = tan^{-1} x$.

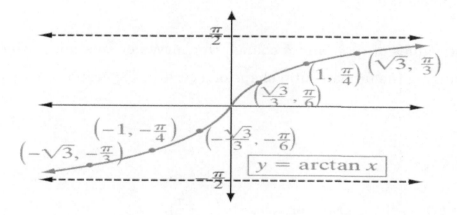

## Example:

Draw the function $y = -2\,tan^{-1}(x-1)$.

**Solution**: The graph of $y = -2\,tan^{-1}(x-1)$ is the graph of $y = tan^{-1} x$ but the phase shift is 1.

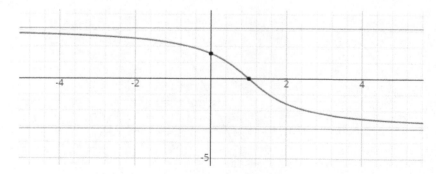

# Sketching Trigonometric Graphs

- For $y = a \cos b(x + c)$ and $y = a \sin b(x + c)$:

  ❖ amplitude $= |a|$

  ❖ number of cycles in a $2\pi$ interval $= |b|$

  ❖ period of the graph $= \dfrac{2\pi}{|b|}$

  ❖ phase shift $= -c$

- The values of $a$, $b$, and $c$ change the curves of sine and cosine without changing the fundamental shape of a cycle of the graph.

## Example:

Draw two cycles of the graph of $y = 2 \sin\left(x - \frac{\pi}{4}\right)$.

**Solution**: For $y = 2 \sin\left(x - \frac{\pi}{4}\right)$, $a = 2$, $b = 1$, $c = -\frac{\pi}{4}$. So, one cycle starts at $x = \frac{\pi}{4}$. There is a complete cycle in the interval $2\pi$, which is from $\frac{\pi}{4}$ to $\frac{9\pi}{4}$. Divide this interval into four equal intervals and draw one cycle of the sine curve with a maximum of 2 and a minimum of $-2$.

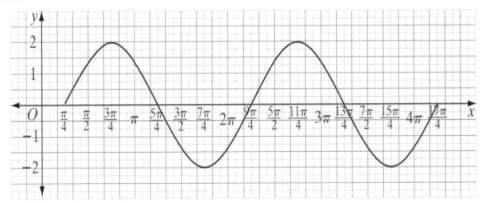

# Chapter 16: Practices

✎ **Graph the following functions.**

1) $y = 2 \sin 2x$

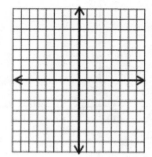

2) $y = -3 \sin x$

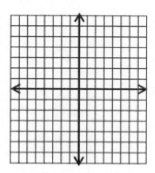

✎ **Graph the following functions.**

3) $y = -2 \cos x$

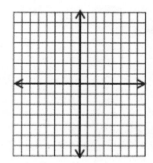

4) $y = 3 \cos 2x - 2$

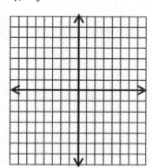

✎ **Determine the amplitude, the period, and the phase shift of:**

5) $y = \sin\left(x - \frac{\pi}{4}\right) - 2$
   Amplitude: ____
   Period: ____
   Phase shift: ____

7) $y = -2 \sin\left(\frac{2}{3}x - \frac{\pi}{3}\right)$
   Amplitude: ____
   Period: ____
   Phase shift: ____

6) $y = 3 \cos\left(2x - \frac{\pi}{6}\right)$
   Amplitude: ____
   Period: ____
   Phase shift: ____

8) $y = \frac{2}{3} \cos\left(2x + \frac{\pi}{3}\right) - 2$
   Amplitude: ____
   Period: ____
   Phase shift: ____

Effortless
Math
Education

✎ **Determine the equation of the graph below.**

9) _____

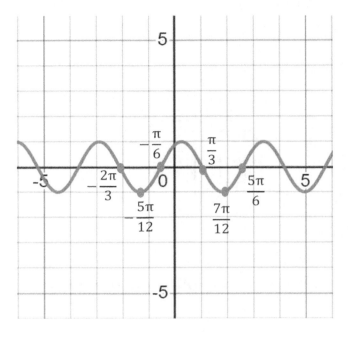

✎ **Determine the equation of the graph below.**

10) _____

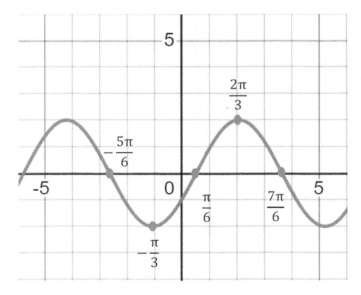

✏ **Draw the graph of equations.**

11) $y = \tan\left(x - \frac{\pi}{4}\right)$

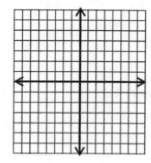

13) $y = 4\sin^{-1}(x + 4)$

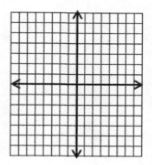

12) $y = 2\csc(3x)$

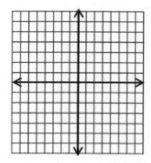

14) $y = 2\cos^{-1}\left(x - \frac{1}{2}\right)$

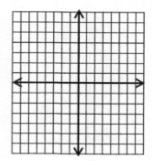

✏ **Draw two cycles of the graph.**

15) $y = 3\cos\left(x - \frac{\pi}{2}\right)$

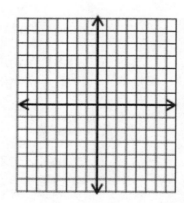

**Effortless Math Education**

# Chapter 16: Answers

1)

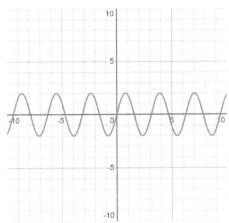

3)

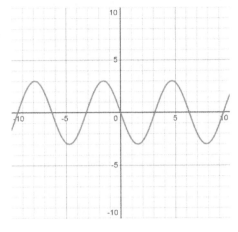

2)

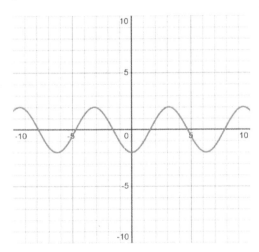

4)

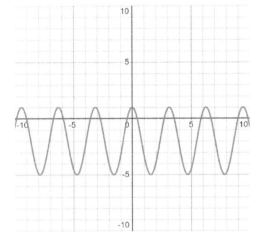

5)  Amplitude: 1
    Period: $2\pi$
    Phase shift: $\frac{\pi}{4}$

6)  Amplitude: 3
    Period: $\pi$
    Phase shift: $\frac{\pi}{12}$

7)  Amplitude: 2
    Period: $3\pi$
    Phase shift: $\frac{\pi}{2}$

8)  Amplitude: $\frac{2}{3}$
    Period: $\pi$
    Phase shift: $-\frac{\pi}{6}$

9) $y = sin\left(2x + \frac{\pi}{3}\right)$

10) $y = 2\,cos\left(x - \frac{2\pi}{3}\right)$

**Effortless**
**Math**
**Education**

11)

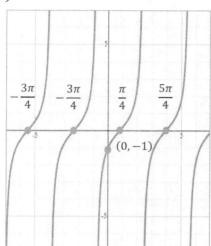

12)

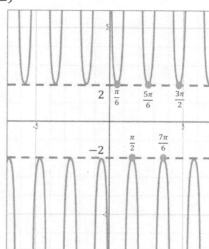

13)

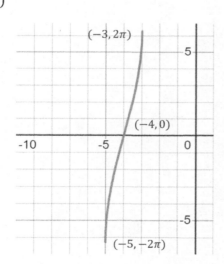

14)

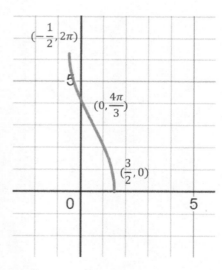

15)

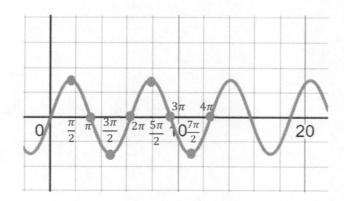

# CHAPTER

# 17 Statistics

Math topics that you'll learn in this chapter:

- ☑ Frequency and Histograms
- ☑ Box-and-Whisker Plots
- ☑ Measures of Dispersion
- ☑ Organizing Data
- ☑ Data Distribution
- ☑ Central Limit Theorem and Standard Error

247

# Frequency and Histograms

- A graph with vertical columns that illustrate the frequency of a data point or range of data points happening in a set of data is named a frequency histogram. The data can be arranged into a frequency histogram to help us better analyze the information. In fact, a frequency histogram is a helpful tool that we can use to visualize a data set and make it better to comprehend.

- In a frequency histogram, the width and position of rectangles can be used to show the various classes, and the heights of those rectangles imply the frequency with which data fell into the related class. In a frequency histogram, there is no space between the bars.

- The frequency histogram consists of a title, an $x-$axis, a $y-$axis, and bars. The $x-$axis lists all the mutually exclusive results from the data set, and each of them comes with its own bar. The $y-$axis has a scale that presents the frequency of the outcome in the data set. The height of the bar aligns with the well-suited number on the scale on the $y-$axis to indicate the frequency of that particular outcome happening in the data set.

## Example:

In a survey, participants of different ages were asked about the role of the media in changing people's views. The following table shows the number of participants in different age groups. Use this frequency table to make a histogram.

| Age | frequency |
|---|---|
| 20 − 30 | 15 |
| 30 − 40 | 10 |
| 40 − 50 | 10 |
| 50 − 60 | 5 |
| 60 − 70 | 20 |

**Solution**: As we mentioned a frequency histogram consists of an $x-$ axis, a $y-$axis, and bars. Here, the $x-$axis can be labeled "Age" and the $y-$axis can be labeled "frequency". The height of the bar in the frequency histogram shows the frequency in the data set.

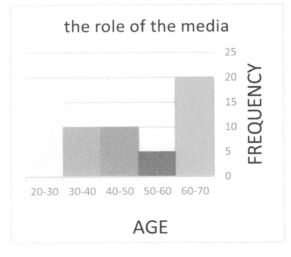

# Box-and-Whisker Plots

- The Box and Whisker plot consists of the following parts:
  - **Median:** This shows the middle point of the data and is represented by a line that divides the box into two parts. Half of the values are greater than or equal to this value and half are less.
  - **First Quartile** ($Q1$): This indicator shows the value that 25% of the data is smaller than. This value makes the left body of the box by a vertical line.

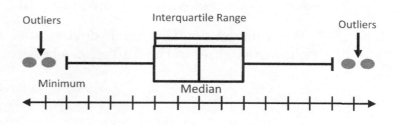

  - **Third Quartile** ($Q3$): This indicator shows the value that 75% of the data is smaller than. This indicator is also used to display the right side of the box.
  - **Interquartile Range** ($IQR$): The distance between the first and third quartiles is shown by this index. The length of the other sides of the box is determined by this index.
  - **Whiskers:** These lines fill the gap between the first quartile and the lowest value as well as the highest value.
  - **Maximum:** In this plot, the maximum is the largest value that is at most 1.5 times the interquartile range away from the third quartile.
  - **Minimum:** In this plot, the minimum is the lowest value that is at most 1.5 times the interquartile range away from the first quartile.
  - **Outlier data:** Data that is smaller than the minimum or larger than the maximum is considered an outlier.

## Example:

According to the following data, draw the related box-and-whisker plots.

$$10, 15, 18, 21, 22, 24, 25, 29, 31, 35$$

*Solution*: According to the data, the median is equal to the average of the two numbers 22 and 24: $\frac{22+24}{2} = 23$. In this boxplot, the minimum is equal to 10 and the maximum is equal to 35. Also, $Q1 = 18$, $Q3 = 29$, and $IQR = 29 - 18 = 11$.

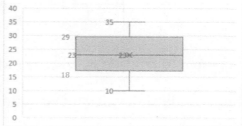

# Measures of Dispersion

- Measures of dispersion can be defined as positive real numbers that measure how homogeneous or heterogeneous the given data is. The value of a measure of the dispersion of a data set is 0 if the points are the same.

- Measures of Dispersion for a data set contain range, interquartile range, mean deviation, variance and standard deviation.

- Mean deviation: The mean deviation of $x_1$, $x_2$, $\cdots$, $x_n$ is the mean of the absolute values of the differences between the data values and the mean, $\bar{x}$.
$$\text{Mean deviation} = \frac{1}{n}\sum_{i=1}^{n}|x_i - \bar{x}|$$

- Variance and standard deviation: If a data set has $n$ data values $x_1$, $x_2$, $\cdots$, $x_n$ and mean $\bar{x}$, then the variance and standard deviation of the data are defined as follows:
$$\text{Variance: } \sigma^2 = \frac{1}{n}\sum_{i=1}^{n}(x_i - \bar{x})^2$$
$$\text{Standard deviation: } \sigma = \sqrt{\sigma^2}$$

## Example:

Find the mean deviation and standard deviation of the data set table:

| Data | 38 | 40 | 42 | 27 | 46 | 39 | 41 |
|------|----|----|----|----|----|----|----|

***Solution:*** For mean deviation $= \frac{1}{n}\sum_{i=1}^{n}|x_i - \bar{x}|$, first compute for each observation $x$ it's deviation $x_i - \bar{x}$ from the mean. Since the mean of the data is:

$\bar{x} = \frac{38+40+42+27+46+39+41}{7} = \frac{273}{7} = 39$,

We obtain the numbers displayed in the other line of the supplied table.

| $i$ | 1 | 2 | 3 | 4 | 5 | 6 | 7 |
|-----|----|----|----|-----|----|----|----|
| $x_i$ | 38 | 40 | 42 | 27 | 46 | 39 | 41 |
| $x_i - \bar{x}$ | $-1$ | 1 | 3 | $-12$ | 7 | 0 | 2 |

So, mean deviation $= \frac{1}{7}\sum_{i=1}^{7}|x_i - \bar{x}| = \frac{1+1+3+12+7+0+2}{7} = \frac{26}{7} \cong 3.71$.

For standard deviation use defining formulas $\sigma = \sqrt{\sigma^2}$, and $\sigma^2 = \frac{1}{n}\sum_{i=1}^{n}(x_i - \bar{x})^2$.

Then:   $\sigma^2 = \frac{1}{7}\sum_{i=1}^{7}(x_i - \bar{x})^2 = \frac{(-1)^2+(1)^2+(3)^2+(-12)^2+(7)^2+(0)^2+(2)^2}{7} = \frac{208}{7} \cong$ $29.71 \rightarrow \sigma \cong 5.45$.

# Organizing Data

- Data organization is a kind of categorization and classification of information to make it more utilizable. Organizing and presenting data is necessary for statistics. After data is gathered, the data may not be meaningful and reasonable when you look at it. That's why it's necessary to arrange and demonstrate the data using tables and charts. You should organize your data in the most reasonable way, so you can find the data you are looking for simply.
- There are two types of data: Categorical or Qualitative data and Quantitative data.
- Categorical or Qualitative data is a kind of data that after being recorded you can't easily recognize with the real numbers. Examples of Categorical data include the colors of something, the size of something as small, medium, large, and gender (male, female). To demonstrate this type of information you can use a bar graph or pie graph.
- Quantitative data is a kind of data that after being recorded can be easily recognized with real numbers. Examples of Quantitative data are age, height, and weight. The fact that you can identify Quantitative data with real numbers makes it easy to organize, compare, and communicate this data. you are also able to combine this data using algebraic operations.

## Examples:

**Example 1.** In research, various types of blood groups ($A$, $B$, $AB$, and $O$) are studied. If we want to organize the data in this research, in which group of data organization are they placed?
*Solution*: The types of blood groups are qualitative data because you can't easily recognize them with real numbers.

**Example 2.** In a survey, participants were asked questions about their height. Now we have a data set containing the heights of 100 randomly chosen 30 years old boys (in $cm$). The data from this survey falls into which category of data organization?
*Solution*: The data obtained from the survey about the heights of 100 randomly chosen 30 years old boys (in $cm$) is included in the category of quantitative data because after being recorded it can be easily recognized with the real numbers.

**Example 3.** If you want to measure the amount of air pollutants during a month in parts per million ($ppm$), the data from this measurement will be classified in which category of data organization?
*Solution*: The amounts of air pollutants during a month in parts per million ($ppm$) are placed in the category of quantitative data because in this data set, you are dealing with real numbers.

bit.ly/3QQk3A6

Find more at

# Data Distribution

- Data distribution is a performance that specifies all possible values for a variable and also quantifies the probability of how they typically occur (or relative frequency). Distributions are considered for any population that features a scattering of information.
- Data distributions are often utilized in statistics. The data distribution method is used for the organization of the raw data into visual and graphical methods like histograms, box plots, etc. and provides helpful info.
- Data can be "distributed" in different ways. There are many cases where the data tends to cluster around the mean without being skewed to the left or right, approaching the so-called "normal distribution."
- The normal distribution has the following properties: $mean = median = mode$, symmetry in the middle, 50% of the values are smaller than the mean and 50% of the values are larger than the mean. The normal distribution is a continuous distribution, that is, it is not discrete and can take any value, and its shape is like a bell (so-called Bell Shape) and is symmetrical around its mean.
- There are different ways and methods to evaluate the normality of data. A handy tool for this is the Histogram chart. The histogram is a useful visual tool to see the shape of data distribution. Based on that, it can be determined to some extent whether the data is normal or almost normal. A stem-and-leaf plot is similar to a histogram, except that here the original data can be reconstructed. A box plot can also help in this direction. Based on the box diagram, we will find out whether the box shape is divided equally into quartiles or unequally spaced between them. Other various types of graphs that can be used for representing data are: Bar graphs, Scatter plots, Line graphs, Area plots, and Pie charts.

## Examples:

**Example 1.** Which graph can be more suitable to display the data distribution of Outlier data?

*Solution*: Considering that Minimum, First Quartile-$Q1$, Median, Third Quartile-$Q3$, and Maximum are specified in the box plot, this graph can give us better information about the existence of outlier data and determine their value.

**Example 2.** In a data set, the data distribution is such that the mean of the data is equal to 25 and the mode of the data is equal to 10. Determine whether this data distribution is the normal distribution or not.

*Solution*: In the normal distribution mean, median, and mode are equal to each other. So, this data distribution is not a normal distribution.

# Central Limit Theorem and Standard Error

- The standard error expresses the difference between the population mean and the mean of each random sample of that population. The formula is as follows:

$$SE = \frac{\sigma}{\sqrt{n}}$$

  Where, $\sigma$ is the standard deviation and $n$ is the sample size.

- The central limit theorem states that when the large sample size ($n \geq 30$) has a finite variance ($\sigma < \infty$), the samples will be normally distributed and the mean of samples will be approximately equal to the mean of the whole population. Its formula is as follows:

$$Z = \frac{\bar{x} - \mu}{\frac{\sigma}{\sqrt{n}}}$$

  Where, $\bar{x}$ is the simple mean and $\mu$ is the population mean.

## Examples:

**Example 1.** Find the standard error of given observations: 38, 40, 42, 27, 46, 39, 41.

*Solution:* Considering the table, the sample size is $n = 7$. The sample mean is equal to: $\bar{x} = \frac{38+40+42+27+46+39+41}{7} = 39$. By the formula of standard error, we know that:

$SE = \frac{\sigma}{\sqrt{n}}$. Now, find the standard deviation $\sigma = \sqrt{\sigma^2}$, by using the formula of variance:

$\sigma^2 = \frac{1}{n}\sum_{i=1}^{n}(x_i - \bar{x})^2 \rightarrow \sigma^2 \cong 29.71$. Therefore, the standard deviation is $\sigma \cong 5.45$. Finally,

$SE = \frac{\sigma}{\sqrt{n}} \rightarrow SE = \frac{5.45}{\sqrt{7}} \cong 2.06$.

**Example 2.** If the lamps produced in a factory have a lifetime with an average of 53 months and a standard deviation of 6 months. What is the probability of a sample of 36 lamps with a lifetime of less than 52 months?

*Solution:* Considering the information of the problem, the population mean is $\mu = 53$, and the population standard deviation is $\sigma = 6$, and the sample size is $n = 36$. Now, by using the central limit formula: $Z = \frac{\bar{x}-\mu}{\frac{\sigma}{\sqrt{n}}} \rightarrow Z = \frac{52-53}{\frac{6}{\sqrt{36}}} = -1$.

Therefore, by using the $z-$table, we have: $P(Z < -1) = 15.871\%$.

bit.ly/3iUwK07

# Chapter 17: Practices

The graph below shows the distribution of scores of 40 students on a mathematics test.

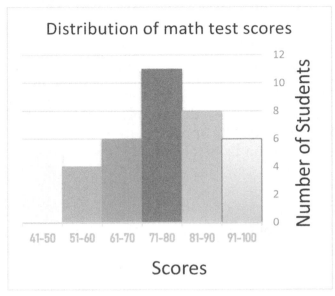

1) Complete the frequency table below using the data in the frequency histogram shown.

| Scores | Frequency |
|---|---|
| 41 − 50 | |
| 51 − 60 | |
| 61 − 70 | |
| 71 − 80 | |
| 81 − 90 | |
| 91 − 100 | |

**According to the following data, draw the related box-and-whisker plots.**

2) 9, 12, 15, 18, 20, 25, 29, 30, 34, 35, 38, 40

**Find the mean deviation and standard deviation of the data set table.**

3) Standard Deviation: _____

Mean Absolute Deviation: _____

| Data | 22 | 28 | 18 | 32 | 16 | 35 | 43 | 40 |
|------|----|----|----|----|----|----|----|----|

4) Standard Deviation: _____

Mean Absolute Deviation: _____

| Data | 55 | 46 | 49 | 58 | 60 | 43 | 62 | 70 |
|------|----|----|----|----|----|----|----|----|

**Solve.**

5) Suppose that total of 45 horse riders participated in the race and their records were registered. The recorded points are normally distributed with a mean point of 76 and a standard deviation of 5.6 points. How many points will fall between 59.2 and 92.8?

**Find the standard error of given observations.**

6) 33, 35, 37, 22, 41, 34, 36.

7) 17, 23, 25, 15, 31, 36, 29.

8) 44, 48, 28, 36, 24, 32, 41.

**Effortless Math Education**

# Chapter 17: Answers

1)

| Scores | Frequency |
|--------|-----------|
| $41 - 50$ | 5 |
| $51 - 60$ | 4 |
| $61 - 70$ | 6 |
| $71 - 80$ | 11 |
| $81 - 90$ | 8 |
| $91 - 100$ | 6 |

2)

Maximum Number: 40

Minimum Number: 9

Third Quartile: 34.5

First Quartile: 16.5

Median: 27

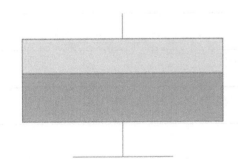

3) Standard Deviation: 9.36

Mean Absolute Deviation: 8.25

4) Standard Deviation: 8.42

Mean Absolute Deviation: 7.125

5) 99.73%

6) standard error: 2.06

7) standard error: 2.64

8) standard error: 3.05

# 18 Probability

Math topics that you'll learn in this chapter:

- ☑ Independent and Dependent Events
- ☑ Compound Events
- ☑ Conditional and Binomial Probabilities
- ☑ Theoretical Probability
- ☑ Experimental Probability

257

# Independent and Dependent Events

- Dependent events affect the likelihood of other events -or their likelihood of happening is influenced by different events.

- Independent events don't have an effect on each other and don't increase or decrease the likelihood of another event happening.

- To identify dependent events or independent events, you can follow these steps:

  - 1st step: Determine whether it's possible for the events to occur in any order or not. If the answer is yes, go to the 2nd step.
  - 2nd step: Determine if one event influences the result of the other event If the answer is yes, go to the 3rd step.
  - 3rd step: If the event is independent, use the formula of independent events and find the answer: $P(A \cap B) = P(A) \cdot P(B)$.
  - 4th step: If the event is dependent, use the formula of dependent events and find the answer: $P(B \ and \ A) = P(A) \times P(B \ after \ A)$.

## Examples:

**Example 1.** In a factory, out of 12 products, 2 of them are defective. If the manufacturer chooses 2 products, what is the probability that both are defective?

***Solution***: The probability that the manufacturer selects a defective product is equal to the number of defective products divided by the total number of products: $\frac{2}{12}$. Once the manufacturer selects a product, only 11 products remain. There is also one less defective product because the manufacturer is not going to choose the same defective product twice. Therefore, the probability that the manufacturer chooses the second defective product is equal to: $\frac{1}{11}$. Therefore, the probability that the manufacturer chooses products in such a way that both are defective is: $\frac{2}{12} \times \frac{1}{11} = \frac{2}{132} = \frac{1}{66}$.

**Example 2.** In a survey, it was found that 4 out of 8 people read books in their free time. If 3 people are randomly selected with replacements, what is the probability that all 3 people read books in their free time?

***Solution***: If 3 people are randomly selected with replacements who read books, then the probability that all 3 people read books is: $P(\text{person 1 reads books}) = \frac{4}{8}$,

$P(\text{person 2 reads books}) = \frac{4}{8}$, $P(\text{person 3 reads books}) = \frac{4}{8}$.

$P(\text{person 1 and person 2 and person 3 read books}) = \frac{4}{8} \times \frac{4}{8} \times \frac{4}{8} = \frac{1}{8}$.

# Compound Events

- The compound probability of compound events is defined as the possibility of 2 or more independent events happening together. An independent event is one whose outcome isn't influenced by the result of other events.

- Compound probability can be determined for 2 sorts of compound events: mutually exclusive compound events and mutually inclusive compound events. The formulas to determine the compound probabilities for each sort of event are different. A mutually inclusive event is a situation where one event can occur with the other, whereas a mutually exclusive event is when 2 events cannot happen at the same time. The compound probability will always be between zero and one.

- To find a compound probability you can follow these steps:
  - 1st step: Read the question and determine if the compound event is a mutually exclusive compound event or a mutually inclusive compound event.
  - 2nd step: List the given information and probabilities.
  - 3rd step: Choose the right compound probability formula: The mutually exclusive events compound probability formula is $P(A \ or \ B) = P(A) + P(B)$ and the mutually inclusive events compound probability formula is: $P(A \ or \ B) = P(A) + P(B) - P(A \ and \ B)$.
  - 4th step: Find the value of $P(A)$ and $P(B)$.
  - 5th step: Put the values into the respective formula to find the answer.

## Examples:

**Example 1.** We have thrown a coin 6 times. What is the probability that we obtain at least one tail?

**Solution**: First, find the total number of outcomes. we have thrown a coin 6 times so the total number of outcomes is equal to $2^6$: $2 \times 2 \times 2 \times 2 \times 2 \times 2 = 2^6 = 64$. Now we should find the desired outcomes (at least one tail): 63. The final step is to find the required probability: $\frac{63}{64}$.

**Example 2.** If we roll two dice, what is the probability that their sum is 3 or less?

**Solution**: When we roll a dice, for each roll there are 6 possible outcomes. So, the total number of outcomes is equal to $6^2$: $6 \times 6 = 6^2 = 36$. In the next step we should find favorable outcomes: 3. The final step is to find the required probability: $\frac{3}{36} = \frac{1}{12}$.

bit.ly/3D0cEZl

Find more at

# Conditional and Binomial Probabilities

- The conditional probability for events $A$ and $B$:

$$P(A) = \frac{P(A \text{ and } B)}{P(A)}$$

Where $P(A) \neq 0$, which can be written as follows: $P(A) = \frac{P(A \cap B)}{P(A)}$.

- Binomial probability is called the probability of exactly $x$ successes on $n$ repeated trials in an experiment which has two possible outcomes, (Where $0 \leq x \leq n$, $p$ is the probability of success, and $1 - p$ is the probability of failure) and is obtained by the following formula:

$$P = {}_nC_x p^x (1 - p)^{n-x}$$

## Examples:

**Example 1.** A bag contains 3 black, 5 grey, and 8 white marbles. Two marbles are randomly selected. Find the probability that the second marble is black given that the first marble is grey. (Assume that the first marble is not replaced.)

**Solution:** Use the conditional probability formula: $P(A) = \frac{P(A \text{ and } B)}{P(A)}$.

The probability of selecting a grey marble: $P(grey) = \frac{5}{3+5+8} = \frac{5}{16}$.

The probability of selecting a black marble: $P(black) = \frac{3}{3+4+8} = \frac{3}{15}$.

The probability of selecting a grey marble: $P(grey \text{ and } black) = \frac{5}{16} \times \frac{3}{15} = \frac{1}{16}$.

Therefore, $P(black) = \frac{P(grey \text{ and } black)}{P(grey)} = \frac{\frac{1}{16}}{\frac{5}{16}} = \frac{1}{5}$.

**Example 2.** What is the probability of getting 4 tails when you toss a coin 7 times?

**Solution:** Considering that there are two outcomes for a coin toss experiment, and assuming equal probability in the coin toss, i.e., the probability is $\frac{1}{2}$. Let $n = 7$ be the number of repeated trials, $x = 4$ the number of successful trials. Use the formula for binomial probability $P = {}_nC_x p^x (1 - p)^{n-x}$. Therefore,

$$P = {}_7C_4 \left(\frac{1}{2}\right)^4 \left(1 - \frac{1}{2}\right)^{7-4} \rightarrow P = \binom{7}{4}\left(\frac{1}{2}\right)^4\left(\frac{1}{2}\right)^3 \rightarrow P = \frac{7!}{4!(7-4)!}\left(\frac{1}{2}\right)^7 \rightarrow P =$$

$35 \times \frac{1}{128}$. By simplifying: $P \cong 0.27$.

# Theoretical Probability

- Theoretical probability is an approach in probability theory and its usage is for the calculation of the probability of a particular event occurrence. Probability theory is a branch of math that's involved with finding the chance of the occurrence of a random event. It tells us about what ought to occur in a perfect situation without conducting any experiments.

-  The probability that an event can occur lies between zero and one. If the probability is nearer to zero, it means that the event is less likely to happen. Similarly, if the probability is nearer to one it implies that the event has more of a chance of happening.

- To find the theoretical probability of an event, divide the number of favorable outcomes by the number of total outcomes.

## Examples:

**Example 1.** The letters of the word "PROBABILITY" are put in a bag. What is the probability of pulling out the letter "B" from the bag?

***Solution***: First count the total number of letters in the word "PROBABILITY": there are 11 letters in this word. Now count the number of the letter "B" in this word: there are 2 letters "B" in this word. The theoretical probability of an event $= \frac{The\ number\ of\ favorable\ outcomes}{The\ number\ of\ total\ outcomes} = \frac{2}{11}$.

**Example 2.** We have written the numbers 0, 2, 8, 7, 9, and 10 separately on 6 cards and put them in a bag. What is the probability that the number we will pick out is greater than 5?

***Solution***: We know the total number of cards is 6. Now count the number of the cards with numbers greater than 5: there are 4 cards with numbers greater than 5. The theoretical probability of an event $= \frac{The\ number\ of\ favorable\ outcomes}{The\ number\ of\ total\ outcomes} = \frac{4}{6} = \frac{2}{3}$.

**Example 3.** The Environmental Protection Organization predicts that there is a 30% chance that half of the world's vegetation will disappear in the next 20 years. What is the probability that this will not happen?

***Solution***: $p(event) + p(complement\ of\ event) = 100\%$ or 1

$\rightarrow 30\% + p(complement\ of\ event) = 100\%$

$\rightarrow p(complement\ of\ event) = 100\% - 30\% = 70\% \rightarrow \frac{70}{100} = \frac{7}{10} = 0.7.$

bit.ly/3kjxiNm

Find more at

# Experimental Probability

- The chance or occurrence of a specific event is termed its probability. The value of a probability lies between zero and one. It means if it's an impossible event, the probability is zero and if it's a definite event, the probability is one.

- Experimental probability is a kind of probability that occurs based on a series of experiments. A random experiment is completed and repeated over and over to see their likelihood and every single repetition is considered a trial. The experiment is conducted to find the possibility of an occurrence occurring or not occurring. It could be coin-tossing, die-rolling, or spinner-rotating.

- In math terms, the probability of an event = the number of times an occurrence occurred ÷ the entire number of trials.

- To find the experimental probability you can follow these steps:
  - 1st step: Conduct an experiment and pay attention to the number of times the event happens and also the total number of trials.
  - 2nd step: Divide these 2 numbers to get the Experimental Probability.

## Examples:

**Example 1.** A bookstore recently sold 12 books, 6 of which were novels. What is the experimental probability that the next book will be a novel?

*Solution*: We know that the experimental probability of an event = The number of times an occurrence occurred ÷ The entire number of trials. We also know that out of the 12 books sold, 6 were novels. Therefore, the experimental probability that the next book will be a novel is $= \frac{6}{12}$. Write the fraction in the simplest form: $\frac{1}{2}$.

**Example 2.** A manufacturer of smartwatches, after testing 100 smartwatches, finds that 95 smartwatches are not defective. What is the experimental probability that a smartwatch chosen at random has no defects?

*Solution*: We know that the experimental probability of an event = The number of times an occurrence occurred ÷ The entire number of trials.

So, the experimental probability that a smartwatch chosen at random has no defects $= \frac{95}{100}$. Write the fraction in the simplest form: $\frac{19}{20}$.

# Chapter 18: Practices

✎ **Solve.**

1) A fair die is rolled three times. What is the probability that at least one roll results in a 6?

2) What is the probability of rolling a number less than 3 when a fair 6 −sided die is rolled?

3) What is the probability of a leap year having 52 Sundays?

4) In a bag, there are 3 blue, 5 red, and 8 white marbles. Two marbles are drawn without replacement. If the first marble drawn is red, what is the probability that the second marble drawn is blue?

5) A jar contains 7 red candies, 9 blue candies, and 14 green candies. Two candies are chosen at random without replacement. Given that the first candy is red, what is the probability that the second candy is also red?

6) A coin has been tossed 5 times. What is the probability that at least one toss results in a 1?

7) Three cards are drawn at random from a deck of 52 cards. What is the probability that at least one of the cards drawn is an Ace?

8) A weighted die is rolled. The probability of rolling an even number is $\frac{3}{5}$, and the probability of rolling an odd number is $\frac{2}{5}$. What is the probability of rolling three odd numbers in a row, followed by one even number?

**Effortless Math Education**

9) A biased die is thrown. The chance of landing on an even number is $\frac{4}{7}$, while the chance of landing on an odd number is $\frac{3}{7}$. What is the probability of achieving two successive odd numbers and then one even number?

10) In a pouch with 5 green and 7 red marbles, what is the likelihood of first picking a red marble followed by a green marble, given that the initially selected marble is not returned to the pouch?

11) What is the likelihood of obtaining 2 heads when flipping a coin 6 times?

12) In a workshop, among 10 items, 3 of them are faulty. If the producer selects 2 items, what is the likelihood that both are faulty?

# Chapter 18: Answers

1) The answer is $\frac{91}{216}$.

   The probability of not getting a 6 in one roll is $\frac{5}{6}$, so the probability of not getting a 6 in three rolls is $(\frac{5}{6})^3 = \frac{125}{216}$.

   Therefore, the probability of getting at least one 6:

   $$P = 1 - \frac{125}{216} = \frac{91}{216}$$

2) The answer is $\frac{1}{3}$.

   There are two possible outcomes that satisfy the condition (rolling a 1 or 2), and there are a total of six possible outcomes (rolling any number from 1 to 6). Therefore, the probability of rolling a number less than 3 is:

   $$\frac{2 \ favorable \ outcomes}{6 \ possible \ outcomes} = \frac{1}{3}$$

3) The answer is $\frac{5}{7}$.

   The total number of events is: 7 (number of days in a week).

   A leap year has 366 days, which is equal to 52 weeks and 2 days. And two days could be any days amongst Sunday, Monday, Tuesday, Wednesday, Thursday, Friday or Saturday. So, $7 - 2 = 5$ is number of favorable outcomes.

   Therefore, the probability of a leap year having 52 Sundays is:

   $$\frac{5 \ favorable \ outcomes}{7 \ possible \ outcomes} = \frac{5}{7}$$

**Effortless**

**Math**

**Education**

4) The answer is $\frac{1}{5}$.

The bag has a total $3 + 5 + 8 = 16$ marbles.

There are 5 red marbles in the bag, and if one of them is drawn first, there will be a total of $16 - 1 = 15$ marbles remaining in the bag. And out of these 15 marbles, 3 marbles will be blue (Reducing the number of red marbles does not affect the number of blue marbles).

Therefore, the probability that the second marble drawn is blue, given that the first marble drawn was red, is:

$$\frac{3 \text{ blue marbles}}{15 \text{ total marbles}} = \frac{1}{5}$$

5) The answer is $\frac{6}{29}$.

The jar has a total $7 + 9 + 14 = 30$ candies.

There are 7 red candies in the jar, so if one of them is drawn first, there will be a total of 29 candies remaining in the jar. Out of these 29 candies, there will be 6 red candies remaining.

Therefore, the probability that the second candy drawn is also red:

$$\frac{6 \text{ red candies}}{29 \text{ total candies}} = \frac{6}{29}$$

6) The answer is $\frac{31}{32}$.

First, find the total number of outcomes. we have thrown a coin 5 times so the total number of outcomes is equal to $2^5$:

$$2 \times 2 \times 2 \times 2 \times 2 = 2^5 = 32$$

Now we should find the desired outcomes (at least one tail):

$$32 - 1 = 31$$

The final step is to find the required probability:

$$\frac{31}{32}$$

7) The answer is $\frac{1,201}{5,525}$.

The probability that none of the three cards drawn are Aces can be calculated as follows:

The probability of drawing a non-Ace card on the first draw is $\frac{48}{52}$, since there are 48 non-Ace cards out of 52 cards total. Similarly, the probability of drawing a non-Ace card on the second draw, after one non-Ace card has already been drawn, is $\frac{47}{51}$. Finally, the probability of drawing a non-Ace card on the third draw, after two non-Ace cards have already been drawn, is $\frac{46}{50}$.

Therefore, the probability that all three cards drawn are non-Aces is:

$$\frac{48}{52} \times \frac{47}{51} \times \frac{46}{50} = \frac{103,776}{132600} = \frac{4,324}{5,525}$$

Therefore, the probability of getting at least one Ace:

$$P = 1 - \frac{4,324}{5,525} = \frac{1,201}{5,525}$$

8) The answer is $\frac{24}{625}$.

To find the probability of rolling three odd numbers in a row, followed by one even number, we simply multiply the individual probabilities together:

Probability of rolling an odd number (first roll) $= \frac{2}{5}$

Probability of rolling an odd number (second roll) $= \frac{2}{5}$

Probability of rolling an odd number (third roll) $= \frac{2}{5}$

Probability of rolling an even number (fourth roll) $= \frac{3}{5}$

Total probability $= \frac{2}{5} \times \frac{2}{5} \times \frac{2}{5} \times \frac{3}{5} = \frac{24}{625}$

**Effortless Math Education**

9) The answer is $\frac{36}{343}$.

To find the probability of achieving two successive odd numbers and then one even number when the die is thrown, you simply multiply the probabilities of each individual event:

Probability of two odd numbers in a row: $\frac{3}{7} \times \frac{3}{7}$

Probability of one even number: $\frac{4}{7}$

Now, multiply the probabilities together: $\frac{3}{7} \times \frac{3}{7} \times \frac{4}{7} = \frac{36}{343}$

So, the probability of achieving two successive odd numbers and then one even number is $\frac{36}{343}$.

10) The answer is $\frac{35}{132}$.

To find the probability of first picking a red marble followed by a green marble without replacement, we can use the formula:

$$P(B \ and \ A) = P(A) \times P(B \ after \ A)$$

In this case, event A is picking a red marble, and event B is picking a green marble after a red marble has been picked.

$P(A)$ = Probability of picking a red marble first:

$$P(A) = \frac{(Number \ of \ red \ marbles)}{(Total \ number \ of \ marbles)} = \frac{7}{12}$$

Once a red marble has been picked, there are 11 marbles remaining in the pouch, with 5 of them being green.

$P(B \ after \ A)$ = Probability of picking a green marble after picking a red marble:

$$P(B \ after \ A) = \frac{(Number \ of \ green \ marbles \ remaining)}{(Total \ number \ of \ marbles \ remaining)} = \frac{5}{11}$$

Now we can calculate the probability of both events occurring:

$$P(B \ and \ A) = P(A) \times P(B \ after \ A) = \frac{7}{12} \times \frac{5}{11} = \frac{35}{132}$$

Therefore, the likelihood of first picking a red marble followed by a green marble without replacement is $\frac{35}{132}$.

11) The answer is 0.234375.

To find the probability of obtaining 2 heads when flipping a fair coin 6 times, we can use this probability formula:

$$P = {}_nC_x p^x (1-p)^{n-x}$$

In this case, $n$ is the number of trials (flips), $k$ is the number of successful outcomes (getting heads), and $p$ is the probability of success on a single trial.

$n = 6$ (Flipping the coin 6 times)

$k = 2$ (Wanting to get 2 heads)

$p = 0.5$ (Probability of getting a head in a single flip, since the coin is fair)

First, we need to calculate the binomial coefficient ${}_nC_x = C(n, k)$, which can be found using the formula:

$$C(n, k) = \frac{n!}{(k!\,(n-k)!)} \rightarrow C(6,2) = \frac{6!}{(2!\,(6-2)!)} = \frac{6!}{2! \times 4!} \rightarrow C(6,2) = 15$$

Now, we can calculate the probability:

$$P = C(6,2) \times 0.5^2 \times (1-0.5)^{6-2} \rightarrow P = 15 \times 0.25 \times (0.5)^4 \rightarrow P = 0.234375$$

12) The answer is $\frac{1}{15}$.

To find the probability of selecting both faulty items from a total of 10 items, of which 3 are faulty, we can use combinations:

$$P(A) = \frac{(Number\ of\ favorable\ outcomes)}{(Total\ possible\ outcomes)}$$

The number of ways to choose 2 items from 3 faulty items (favorable outcomes) can be found using combinations:

$$C(n, k) = \frac{n!}{(k!\,(n-k)!)} \rightarrow C(3,2) = \frac{3!}{(2!\,(3-2)!)} \rightarrow C(3,2) = 3$$

Now we need to find the total number of ways to choose 2 items from the 10 items (total possible outcomes):

$$C(10,2) = \frac{10!}{(2!\,(10-2)!)} \rightarrow C(10,2) = 45$$

Now we can find the probability:

$$P(A) = \frac{(Number\ of\ favorable\ outcomes)}{(Total\ possible\ outcomes)} = \frac{3}{45} = \frac{1}{15}$$

**Effortless Math Education**

# Time to Test

## Time to refine your Math skill with a practice test

In this section, there are two complete Algebra 2 Tests. Take these tests to simulate the test day experience. After you've finished, score your test using the answer keys.

## Before You Start

- You'll need a pencil a calculator to take the test.
- For each question, there are five possible answers. Choose which one is best.
- It's okay to guess. There is no penalty for wrong answers.
- Use the answer sheet provided to record your answers.
- **Calculator is permitted for Algebra 2 Test.**
- After you've finished the test, review the answer key to see where you went wrong.

*Good luck!*

# Algebra 2

# Practice Test 1

## 2022 - 2023

Total number of questions: 60

Time: <u>No time limit</u>

Calculator is permitted for Algebra 2 Test.

# Algebra 2 Practice Test 1 Answer Sheet

Remove (or photocopy) this answer sheet and use it to complete the practice test.

| Algebra 2 Practice Test 1 Answer Sheet |
|---|

1  Ⓐ Ⓑ Ⓒ Ⓓ Ⓔ     21  Ⓐ Ⓑ Ⓒ Ⓓ Ⓔ     41  Ⓐ Ⓑ Ⓒ Ⓓ Ⓔ
2  Ⓐ Ⓑ Ⓒ Ⓓ Ⓔ     22  Ⓐ Ⓑ Ⓒ Ⓓ Ⓔ     42  Ⓐ Ⓑ Ⓒ Ⓓ Ⓔ
3  Ⓐ Ⓑ Ⓒ Ⓓ Ⓔ     23  Ⓐ Ⓑ Ⓒ Ⓓ Ⓔ     43  Ⓐ Ⓑ Ⓒ Ⓓ Ⓔ
4  Ⓐ Ⓑ Ⓒ Ⓓ Ⓔ     24  Ⓐ Ⓑ Ⓒ Ⓓ Ⓔ     44  Ⓐ Ⓑ Ⓒ Ⓓ Ⓔ
5  Ⓐ Ⓑ Ⓒ Ⓓ Ⓔ     25  Ⓐ Ⓑ Ⓒ Ⓓ Ⓔ     45  Ⓐ Ⓑ Ⓒ Ⓓ Ⓔ
6  Ⓐ Ⓑ Ⓒ Ⓓ Ⓔ     26  Ⓐ Ⓑ Ⓒ Ⓓ Ⓔ     46  Ⓐ Ⓑ Ⓒ Ⓓ Ⓔ
7  Ⓐ Ⓑ Ⓒ Ⓓ Ⓔ     27  Ⓐ Ⓑ Ⓒ Ⓓ Ⓔ     47  Ⓐ Ⓑ Ⓒ Ⓓ Ⓔ
8  Ⓐ Ⓑ Ⓒ Ⓓ Ⓔ     28  Ⓐ Ⓑ Ⓒ Ⓓ Ⓔ     48  Ⓐ Ⓑ Ⓒ Ⓓ Ⓔ
9  Ⓐ Ⓑ Ⓒ Ⓓ Ⓔ     29  Ⓐ Ⓑ Ⓒ Ⓓ Ⓔ     49  Ⓐ Ⓑ Ⓒ Ⓓ Ⓔ
10 Ⓐ Ⓑ Ⓒ Ⓓ Ⓔ     30  Ⓐ Ⓑ Ⓒ Ⓓ Ⓔ     50  Ⓐ Ⓑ Ⓒ Ⓓ Ⓔ
11 Ⓐ Ⓑ Ⓒ Ⓓ Ⓔ     31  Ⓐ Ⓑ Ⓒ Ⓓ Ⓔ     51  Ⓐ Ⓑ Ⓒ Ⓓ Ⓔ
12 Ⓐ Ⓑ Ⓒ Ⓓ Ⓔ     32  Ⓐ Ⓑ Ⓒ Ⓓ Ⓔ     52  Ⓐ Ⓑ Ⓒ Ⓓ Ⓔ
13 Ⓐ Ⓑ Ⓒ Ⓓ Ⓔ     33  Ⓐ Ⓑ Ⓒ Ⓓ Ⓔ     53  Ⓐ Ⓑ Ⓒ Ⓓ Ⓔ
14 Ⓐ Ⓑ Ⓒ Ⓓ Ⓔ     34  Ⓐ Ⓑ Ⓒ Ⓓ Ⓔ     54  Ⓐ Ⓑ Ⓒ Ⓓ Ⓔ
15 Ⓐ Ⓑ Ⓒ Ⓓ Ⓔ     35  Ⓐ Ⓑ Ⓒ Ⓓ Ⓔ     55  Ⓐ Ⓑ Ⓒ Ⓓ Ⓔ
16 Ⓐ Ⓑ Ⓒ Ⓓ Ⓔ     36  Ⓐ Ⓑ Ⓒ Ⓓ Ⓔ     56  Ⓐ Ⓑ Ⓒ Ⓓ Ⓔ
17 Ⓐ Ⓑ Ⓒ Ⓓ Ⓔ     37  Ⓐ Ⓑ Ⓒ Ⓓ Ⓔ     57  Ⓐ Ⓑ Ⓒ Ⓓ Ⓔ
18 Ⓐ Ⓑ Ⓒ Ⓓ Ⓔ     38  Ⓐ Ⓑ Ⓒ Ⓓ Ⓔ     58  Ⓐ Ⓑ Ⓒ Ⓓ Ⓔ
19 Ⓐ Ⓑ Ⓒ Ⓓ Ⓔ     39  Ⓐ Ⓑ Ⓒ Ⓓ Ⓔ     59  Ⓐ Ⓑ Ⓒ Ⓓ Ⓔ
20 Ⓐ Ⓑ Ⓒ Ⓓ Ⓔ     40  Ⓐ Ⓑ Ⓒ Ⓓ Ⓔ     60  Ⓐ Ⓑ Ⓒ Ⓓ Ⓔ

1) If $A \times B = [-1 \quad 4]$, where $A$ and $B$ are matrices, which are $A$ and $B$?

    A. $A = [2 \quad 1]$ and $B = [-1 \quad 2]$

    B. $A = [2 \quad 1]$ and $B = \begin{bmatrix} 0 & 2 \\ -1 & 0 \end{bmatrix}$

    C. $A = [2 \quad 0]$ and $B = \begin{bmatrix} 1 & 0 \\ 1 & 3 \end{bmatrix}$

    D. $A = \begin{bmatrix} 2 \\ 0 \end{bmatrix}$ and $B = \begin{bmatrix} 1 \\ 3 \end{bmatrix}$

2) In an arithmetic sequence where the 3rd sequence is 13 and the 7th sequence is 25, what is the sum of the 10th sequence of this series?

    A. 171

    B. 205

    C. 308

    D. 265

3) Which statement best describes these two functions?
$$f(x) = 2x^2 - 4x$$
$$g(x) = (x - 1)^2 - 1$$

    A. They have no common points.

    B. They have the same $x$ −intercepts.

    C. The maximum of $f(x)$ is the same as the minimum of $g(x)$.

    D. They have the same minimum.

4) Which expression is another way to write $\sqrt[3]{64x^5}$?

    A. $8x^{\frac{5}{3}}$

    B. $4x^{\frac{5}{3}}$

    C. $4x^{\frac{3}{5}}$

    D. $8x^{\frac{3}{5}}$

5) If $e^{\ln 4} = x$, what is the value of $x$?

    A. $e$

    B. 4

    C. 2

    D. $e^4$

6) What is the simplification of this rational expression: $\frac{5x}{x+3} \div \frac{x}{2x+6} =?$

   A. $\frac{5x^2}{2x^2+12x+18}$

   B. $\frac{5}{2}x$

   C. $9x(x+3)$

   D. $10$

7) What is the range for $f(x) = x^2 + 2$?

   A. $\mathbb{R}$

   B. $(-2, +\infty)$

   C. $(2, +\infty)$

   D. $[2, +\infty)$

8) If $f(x) = \frac{1}{2}x - 1$ and $g(x) = -2$, which graph corresponds to the function of $(f \circ g)(x)$?

   A. $M$

   B. $P$

   C. $O$

   D. $N$

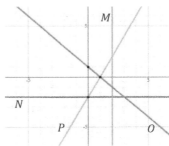

9) If $x = 3\left(log_2 \frac{1}{32}\right)$, what is the value of $x$?

   A. $-15$

   B. $\frac{3}{5}$

   C. $15$

   D. $-\frac{3}{5}$

10) What is the value of $x$ in the following equation?
$$log_4(x+2) - log_4(x-2) = 1$$

   A. $log_4 10$

   B. $\frac{10}{3}$

   C. $10$

   D. $-\frac{10}{3}$

11) If $\tan \theta = \frac{5}{12}$ and $\sin \theta < 0$, then $\cos \theta = ?$

   A. $\frac{12}{13}$

   B. $-\frac{5}{13}$

   C. $-\frac{12}{13}$

   D. $\frac{13}{12}$

12) What is the center and radius of a circle with the following equation?
$$(x - 4)^2 + (y + 7)^2 = 3$$

   A. $(4,7)$ and $\sqrt{3}$

   B. $(4,-7)$ and $3$

   C. $(4,7)$ and $3$

   D. $(4,-7)$ and $\sqrt{3}$

13) Which statement best describes these two functions?
$$f(x) = 2x^2 - x + 3$$
$$g(x) = -x^2 + 2x + 1$$

   A. The maximum of $f(x)$ is less than the minimum of $g(x)$.

   B. The minimum of $f(x)$ is less than the maximum of $g(x)$.

   C. The maximum of $f(x)$ is greater than the minimum of $g(x)$.

   D. The minimum of $f(x)$ is greater than the maximum of $g(x)$.

14) Simplify and express in the form $a + bi$: $\frac{-8i}{12+i}$.

   A. $8 - 96i$

   B. $\frac{8}{145} - \frac{96}{145}i$

   C. $\frac{8}{145} - \frac{96}{145}i$

   D. $-\frac{8}{145} - \frac{96}{145}i$

15) Solve the inequality for $x$. Simplify your answer as much as possible.
$$-6 + 2x > -4$$

   A. $x > -1$

   B. $x < -1$

   C. $x \geq 1$

   D. $x > 1$

$$f(x) = \frac{1}{(x-3)^2 + 4(x-3) + 4}$$

16) For what value of $x$ is the function $f(x)$ above undefinded?

   A. $-1$
   B. $1$
   C. $-3$
   D. $3$

17) What is the matrix $A$, if the determinant of $A$ is equal to $-10$?

   A. $\begin{bmatrix} 2 & 6 \\ 0 & 5 \end{bmatrix}$

   B. $\begin{bmatrix} 2 & 6 & 3 \\ 0 & 1 & 1 \\ 4 & 7 & 4 \end{bmatrix}$

   C. $\begin{bmatrix} 1 & -2 \\ -5 & 0 \end{bmatrix}$

   D. $\begin{bmatrix} 1 & 3 & 0 \\ 0 & 5 & 1 \\ 1 & -1 & 0 \end{bmatrix}$

18) In an arithmetic sequence beginning with 15 and ending with 135, how many integers are divisible by 5?

   A. 25
   B. 24
   C. 26
   D. 23

19) Which function represents this graph?

   A. $y = 2x^2 - 2$
   B. $y = -\frac{1}{2}x^2 + 2$
   C. $y = -2x^2 + 2$
   D. $y = -\frac{1}{2}x^2 - 2$

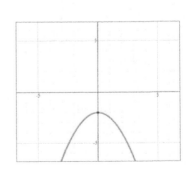

20) Find the function represented by the following graph?

    A. $f(x) = 2x^2 + 4x - 3$
    B. $f(x) = -2x^2 + 4x - 3$
    C. $f(x) = 2x^2 - 4x - 3$
    D. $f(x) = 2x^2 + 4x + 3$

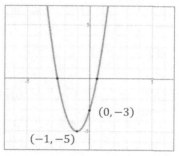

21) What is the graph of linear inequality, $y \leq -2x - 2$?

  A.

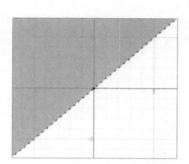

  B.

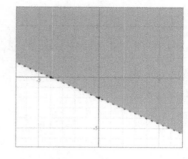

  C.

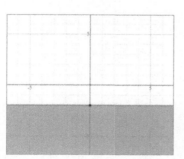

  D.

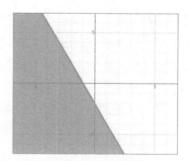

22) What is the equation of the ellipse whose center is at the origin, the major axis has a length of 9 units along the $x$ −axis, and the minor axis has a length of 16 units?

    A. $\frac{x^2}{9} + \frac{y^2}{16} = 1$

    B. $\frac{4x^2}{81} + \frac{y^2}{64} = 1$

    C. $\frac{x^2}{81} + \frac{y^2}{64} = 1$

    D. $\frac{x^2}{(9)^2} + \frac{y^2}{(16)^2} = 1$

23) What is the vertex of the parabola $y = (x - 2)^2 + 3$?

    A. $(-2,3)$
    B. $(-2,-3)$
    C. $(2,3)$
    D. $(2,-3)$

24) What is the equation of a circle with center $(-1,2)$ and perimeter $4\pi$?

    A. $(x + 1)^2 + (y - 2)^2 = 4\pi$
    B. $(x + 1)^2 + (y - 2)^2 = 4$
    C. $(x - 1)^2 + (y + 2)^2 = 2$
    D. $(x - 1)^2 + (y + 2)^2 = 2\pi$

25) What is the simplified expression of $\sqrt[3]{\dfrac{81x^5}{3x^2}}$?

    A. $9x^2$
    B. $3x^2$
    C. $3x$
    D. $9x$

26) If $(f o g)(x) = x - 1$, how might $f(x)$ and $g(x)$ be defined?

    A. $f(x) = x + 1$ and $g(x) = -x - 2$
    B. $f(x) = -x + 1$ and $g(x) = x - 2$
    C. $f(x) = x + 1$ and $g(x) = x + 2$
    D. $f(x) = -x + 1$ and $g(x) = -x + 2$

27) Which equation represents the solution for $x$ in the formula $3(7^{2x}) = 11$?

    A. $x = \frac{1}{2}\left(\frac{\log 11}{\log 3} - \log 7\right)$
    B. $x = \frac{2(\log 11 - \log 3)}{\log 7}$
    C. $x = 2\left(\frac{\log 11}{\log 3} - \log 7\right)$
    D. $x = \frac{\log 11 - \log 3}{2 \log 7}$

28) How many $x$ −intercepts does the graph of $y = \frac{x-1}{1-x^2}$ have?

   A. 0

   B. 1

   C. 2

   D. 3

29) What is the value of $x$ in the geometric sequence $\left\{2, x, \frac{1}{8}, \cdots \right\}$?

   A. 1

   B. $-\frac{1}{2}$

   C. $\frac{1}{4}$

   D. $-2$

30) Find $AC$ in the following triangle. Round your answer to the nearest tenth.

   A. 3.86

   B. 6

   C. 4.59

   D. 5.03

31) Simplify. $(-4 + 9i)(3 + 5i)$

   A. $33 + 7i$

   B. $-57 + 7i$

   C. $57 + 7i$

   D. $-33 + 7i$

32) Find the solution of the following inequality:
$$|x - 10| \leq 4$$

   A. $6 < x \leq 14$

   B. $x \geq 14 \cup x \leq 6$

   C. $6 \leq x \leq 14$

   D. $x \geq 14 \cup x < 6$

33) If $x + sin^2 a + cos^2 a = 3$, then $x =$?

   A. 1

   B. 2

   C. 3

   D. 4

34) Which graph represents the function $-log(1-x)$?

A.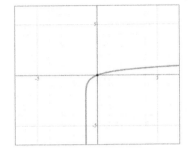

B.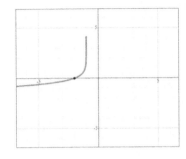

C.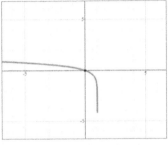

D.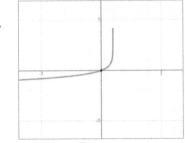

35) Find the positive and negative coterminal angles to angle $\frac{\pi}{2}$.

A. $5\pi, -\frac{3\pi}{2}$

B. $-\frac{5\pi}{2}, -\frac{3\pi}{2}$

C. $\frac{5\pi}{2}, \frac{3\pi}{2}$

D. $\frac{5\pi}{2}, -\frac{3\pi}{2}$

36) Find the center and the radius of a circle with the following equation:

$$x^2 + y^2 - 6x + 4y + 4 = 0$$

A. $(3, -2)$ and 3

B. $(3, -2)$ and $-3$

C. $(-3, 2)$ and 2

D. $(-3, 2)$ and 4

37) Which graph reperesents the inverse of $y = -x$?

A.

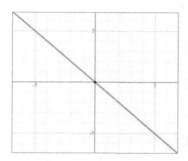

B.

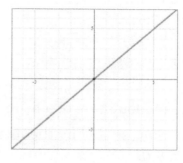

C.

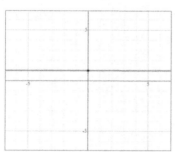

D.

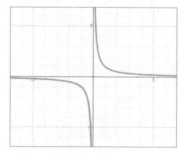

38) If $log_2 x = 5$, then $x =$?

   A. 10

   B. 32

   C. 16

   D. $\frac{5}{2}$

39) What is the equation of the horizontal asymptote of the function $f(x) = \frac{x+3}{x^2+1}$?

   A. $y = 0$

   B. $x = -3$

   C. $x^2 = -1$

   D. No horizontal asymptote

40) Point $A$ lies on the line with equation $y - 3 = 2(x + 5)$. If the $x$ −coordinate of $A$ is 8, what is the $y$ −coordinate of $A$?

   A. 3

   B. 8

   C. 0

   D. 29

41) Which ordered pair $(x, y)$, determines whether it is the solution to the inequality $y < 5x^2 - 2x - 5$?

A. $(-1, 2)$

B. $(3, 6)$

C. $(1, 12)$

D. $(-1, 6)$

42) If the loudness of a drum in concert is represented by $F = 9 \log \frac{S}{10^{-6}}$, where $S$ is represented by the intensity of sound, how loud is the fizz if $S = 10^{-5}$?

A. $9 dB$

B. $18 dB$

C. $81 dB$

D. $-9 dB$

43) If $f(x) = x + 6$ and $g(x) = -x^2 - 2x - 1$, what is $(g - f)(x)$?

A. $x^2 + 3x + 7$

B. $-x^2 + 3x - 7$

C. $-x^2 - 3x - 7$

D. $x^2 - 3x + 7$

44) What is the parent function of this graph?

A. $x^2$

B. $-x^4$

C. $-x^2$

D. $x^4$

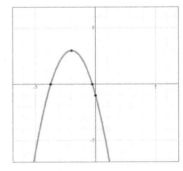

45) What is the value of $\cos 30°$?

A. $\frac{\sqrt{2}}{2}$

B. $\frac{1}{2}$

C. $-\frac{1}{2}$

D. $\frac{\sqrt{3}}{2}$

46) In the standard $(x, y)$ coordinate system plane, find the area of the circle with the following equation.

$$(x + 2)^2 + (y - 4)^2 = 25$$

A. 25

B. $25\pi$

C. $\frac{25}{\pi}$

D. $5\pi$

47) Which function is best represented by this graph?

A. $ln(x)$

B. $ln(x + 1)$

C. $ln(x - 1)$

D. $ln(x) - 1$

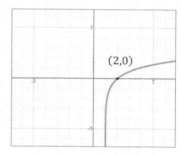

48) If $\begin{bmatrix} 2 & 4 \\ -5 & -1 \\ -2 & -6 \end{bmatrix} + \begin{bmatrix} a & 0 \\ 0 & 7 \\ 3 & 5 \end{bmatrix} = \begin{bmatrix} 1 & 4 \\ -5 & 6 \\ 1 & b \end{bmatrix}$, then $a, b =$?

A. $a = 1$ and $b = -1$

B. $a = -1$ and $b = 1$

C. $a = -1$ and $b = -1$

D. $a = 1$ and $b = 1$

49) Find the equation of the given parabola:

A. $y = x^2 + 4x - 1$

B. $y = -x^2 - 4x$

C. $y = -2x^2 - 4x + 2$

D. $y = -x^2 + 4$

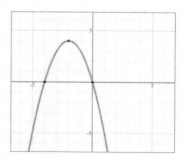

50) Find the value of $x$ in the following triangle. (Round your answer to the closest whole number.)
   A. 140
   B. 45
   C. 44
   D. 24

51) Which is the value of $x$ in the following diagram? (There are 2 supplementary angles in the diagram.)
   A. 54
   B. 61
   C. 45
   D. 90

52) Find side $AC$ in the following triangle. Round your answer to the nearest tenth.
   A. 7.14
   B. 4.6
   C. 3.8
   D. 6

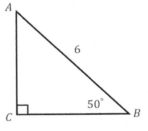

53) What is the graph of $2y > 4x^2$?

A.

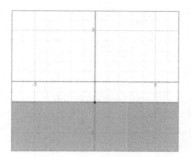

B.

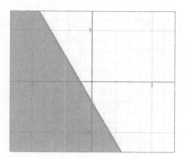

C.

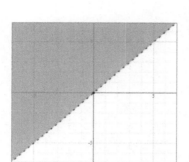

D.

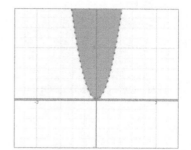

54) Find $tan\frac{2\pi}{3}$?

A. $\sqrt{3}$

B. $-\frac{\sqrt{3}}{3}$

C. $-\sqrt{3}$

D. $\frac{\sqrt{3}}{3}$

55) Solve the following system of equations.

$$x + 2y = 10$$
$$6x - 2y = 18$$

A. (4,3)

B. (3,4)

C. (4,−3)

D. (3,−4)

56) Find the value of $y$ in the following system of equation:
$$3x - 4y = -20$$
$$-x + 2y = 20$$

    A. $-20$
    B. $20$
    C. $40$
    D. $-40$

57) What is the solution of the following inequality?
$$|x - 2| \geq 4$$

    A. $x < -2 \cup x \geq 6$
    B. $x \geq 6$
    C. $x \geq -2 \cup x \geq 6$
    D. $x \leq -2 \cup x \geq 6$

58) Which graph corresponds to $y = (x + 1)^2 - 2$?

    A.                                       B.

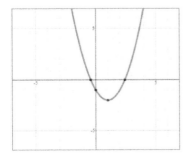 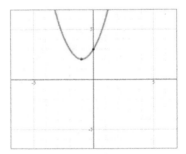

    C.                                       D.

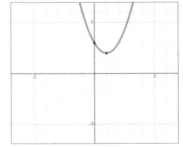

59) What is the least common denominator for $\frac{3x}{x^2-36}$ and $\frac{4}{2x-12}$?

    A. $(x^2 - 6)(2x - 12)$

    B. $x - 6$

    C. $2x^2 - 72$

    D. $2x + 12$

60) If $3x - 2\sin^2 a - 2\cos^2 a = 4$, then $x =$?

    A. 1

    B. 2

    C. 3

    D. 4

**End of Algebra 2 Practice Test 1.**

# Algebra 2
# Practice Test 2

## 2022 - 2023

Total number of questions: 60

Time: <u>No time limit</u>

Calculator is permitted for Algebra 2 Test.

# Algebra 2 Practice Test 2 Answer Sheet

**Remove (or photocopy) this answer sheet and use it to complete the practice test.**

Algebra 2 Practice Test 2 Answer Sheet

| | | | | | |
|---|---|---|---|---|---|
| 1 | Ⓐ Ⓑ Ⓒ Ⓓ Ⓔ | 21 | Ⓐ Ⓑ Ⓒ Ⓓ Ⓔ | 41 | Ⓐ Ⓑ Ⓒ Ⓓ Ⓔ |
| 2 | Ⓐ Ⓑ Ⓒ Ⓓ Ⓔ | 22 | Ⓐ Ⓑ Ⓒ Ⓓ Ⓔ | 42 | Ⓐ Ⓑ Ⓒ Ⓓ Ⓔ |
| 3 | Ⓐ Ⓑ Ⓒ Ⓓ Ⓔ | 23 | Ⓐ Ⓑ Ⓒ Ⓓ Ⓔ | 43 | Ⓐ Ⓑ Ⓒ Ⓓ Ⓔ |
| 4 | Ⓐ Ⓑ Ⓒ Ⓓ Ⓔ | 24 | Ⓐ Ⓑ Ⓒ Ⓓ Ⓔ | 44 | Ⓐ Ⓑ Ⓒ Ⓓ Ⓔ |
| 5 | Ⓐ Ⓑ Ⓒ Ⓓ Ⓔ | 25 | Ⓐ Ⓑ Ⓒ Ⓓ Ⓔ | 45 | Ⓐ Ⓑ Ⓒ Ⓓ Ⓔ |
| 6 | Ⓐ Ⓑ Ⓒ Ⓓ Ⓔ | 26 | Ⓐ Ⓑ Ⓒ Ⓓ Ⓔ | 46 | Ⓐ Ⓑ Ⓒ Ⓓ Ⓔ |
| 7 | Ⓐ Ⓑ Ⓒ Ⓓ Ⓔ | 27 | Ⓐ Ⓑ Ⓒ Ⓓ Ⓔ | 47 | Ⓐ Ⓑ Ⓒ Ⓓ Ⓔ |
| 8 | Ⓐ Ⓑ Ⓒ Ⓓ Ⓔ | 28 | Ⓐ Ⓑ Ⓒ Ⓓ Ⓔ | 48 | Ⓐ Ⓑ Ⓒ Ⓓ Ⓔ |
| 9 | Ⓐ Ⓑ Ⓒ Ⓓ Ⓔ | 29 | Ⓐ Ⓑ Ⓒ Ⓓ Ⓔ | 49 | Ⓐ Ⓑ Ⓒ Ⓓ Ⓔ |
| 10 | Ⓐ Ⓑ Ⓒ Ⓓ Ⓔ | 30 | Ⓐ Ⓑ Ⓒ Ⓓ Ⓔ | 50 | Ⓐ Ⓑ Ⓒ Ⓓ Ⓔ |
| 11 | Ⓐ Ⓑ Ⓒ Ⓓ Ⓔ | 31 | Ⓐ Ⓑ Ⓒ Ⓓ Ⓔ | 51 | Ⓐ Ⓑ Ⓒ Ⓓ Ⓔ |
| 12 | Ⓐ Ⓑ Ⓒ Ⓓ Ⓔ | 32 | Ⓐ Ⓑ Ⓒ Ⓓ Ⓔ | 52 | Ⓐ Ⓑ Ⓒ Ⓓ Ⓔ |
| 13 | Ⓐ Ⓑ Ⓒ Ⓓ Ⓔ | 33 | Ⓐ Ⓑ Ⓒ Ⓓ Ⓔ | 53 | Ⓐ Ⓑ Ⓒ Ⓓ Ⓔ |
| 14 | Ⓐ Ⓑ Ⓒ Ⓓ Ⓔ | 34 | Ⓐ Ⓑ Ⓒ Ⓓ Ⓔ | 54 | Ⓐ Ⓑ Ⓒ Ⓓ Ⓔ |
| 15 | Ⓐ Ⓑ Ⓒ Ⓓ Ⓔ | 35 | Ⓐ Ⓑ Ⓒ Ⓓ Ⓔ | 55 | Ⓐ Ⓑ Ⓒ Ⓓ Ⓔ |
| 16 | Ⓐ Ⓑ Ⓒ Ⓓ Ⓔ | 36 | Ⓐ Ⓑ Ⓒ Ⓓ Ⓔ | 56 | Ⓐ Ⓑ Ⓒ Ⓓ Ⓔ |
| 17 | Ⓐ Ⓑ Ⓒ Ⓓ Ⓔ | 37 | Ⓐ Ⓑ Ⓒ Ⓓ Ⓔ | 57 | Ⓐ Ⓑ Ⓒ Ⓓ Ⓔ |
| 18 | Ⓐ Ⓑ Ⓒ Ⓓ Ⓔ | 38 | Ⓐ Ⓑ Ⓒ Ⓓ Ⓔ | 58 | Ⓐ Ⓑ Ⓒ Ⓓ Ⓔ |
| 19 | Ⓐ Ⓑ Ⓒ Ⓓ Ⓔ | 39 | Ⓐ Ⓑ Ⓒ Ⓓ Ⓔ | 59 | Ⓐ Ⓑ Ⓒ Ⓓ Ⓔ |
| 20 | Ⓐ Ⓑ Ⓒ Ⓓ Ⓔ | 40 | Ⓐ Ⓑ Ⓒ Ⓓ Ⓔ | 60 | Ⓐ Ⓑ Ⓒ Ⓓ Ⓔ |

1) The revenue $R$ at a fast food restaurant is given by the equation $R = \frac{(5000x - x^2)}{200}$, where $x$ is the number of orders. What is the maximum amount of revenue the fast food restaurant can generate?
   A. $200
   B. $5,000
   C. $31,250
   D. Unlimited

2) If the center of a circle is at the point $(-4, 2)$ and its circumference equals to $2\pi$, what is the standard form equation of the circle?
   A. $(x - 4)^2 + (y + 2)^2 = 1$
   B. $(x + 4)^2 + (y - 2)^2 = 2\pi$
   C. $(x + 4)^2 + (y - 2)^2 = 1$
   D. $(x - 4)^2 + (y + 2)^2 = 2$

3) If $\tan x = \frac{8}{15}$, then $\sin x = ?$
   A. $\frac{15}{8}$
   B. $\frac{15}{17}$
   C. $\frac{8}{17}$
   D. $-\frac{15}{17}$

4) Simplify. $\frac{4 - 3i}{-4i}$
   A. $\frac{3}{4} - i$
   B. $-\frac{3}{4} + i$
   C. $-\frac{3}{4} - i$
   D. $\frac{3}{4} + i$

5) If $x \begin{bmatrix} 2 & 0 \\ 0 & 4 \end{bmatrix} = \begin{bmatrix} x + 3y - 5 & 0 \\ 0 & 2y + 10 \end{bmatrix}$, what is the product of $x$ and $y$?

A. 12

B. 7

C. 4

D. 34

6) If $f(x) = -8x^2 + 12x$ and $g(x) = -2x + 3$, then find $\left(\frac{f}{g}\right)(x)$.

A. $-4$

B. $4x$

C. 4

D. $-4x$

7) Find the slope of the line passing through the points $(-4, 3)$ and $(-9, 7)$.

A. $-\frac{4}{5}$

B. $\frac{4}{5}$

C. $\frac{5}{4}$

D. $-\frac{5}{4}$

8) The profit $P$ (in dollars) for a car rental company is given by $P = 120N - 0.2N^2$, where $N$ is the number of cars rented. How many cars have to be rented for the company to maximize profits?

A. 120 cars

B. 300 cars

C. 600 cars

D. The more the better.

9) A ladder leans against a wall forming a $60°$ angle between the ground and the ladder. If the bottom of the ladder is 30 feet away from the wall, how long is the ladder?

   A. $30ft$

   B. $60ft$

   C. $30\sqrt{3}ft$

   D. $30\sqrt{2}ft$

10) If $\frac{5}{x+1} = \frac{x+1}{x^2-1}$, then $x =$?

   A. $-1, 1$

   B. $\frac{3}{2}$

   C. $-1, \frac{3}{2}$

   D. $-1$

11) Find the value of $x$ in this equation: $log(5x + 2) = log(3x - 1)$.

   A. $\frac{11}{2}$

   B. $-\frac{11}{2}$

   C. $-\frac{11}{2}, \frac{11}{2}$

   D. No solution

12) What is the solution of the following inequality?
$$|x - 2| \geq 4$$

   A. $x < -2 \cup x \geq 6$

   B. $x \geq 6$

   C. $x \geq -2 \cup x \geq 6$

   D. $x \leq -2 \cup x \geq 6$

13) If $x$ and $y$ are real numbers, what is the simplified radical form of $(x^3y^5)^{\frac{2}{3}}$?

     A. $x^2y^3\sqrt[3]{y}$

     B. $x^3y^2\sqrt[3]{y}$

     C. $x^2y^3\sqrt{y}$

     D. $x^3y^2\sqrt{y}$

14) What is the vertical asymptote of the graph $y = \frac{5x-6}{3x+4}$?

     A. $x = -\frac{3}{4}$

     B. $y = \frac{3}{4}$

     C. $x = -\frac{4}{3}$

     D. $x = \frac{6}{5}$

15) How many terms are there in a geometric series if the first term is 4, the common ratio is 3, and the sum of the series is 484?

     A. 3

     B. 5

     C. 4

     D. 12

16) Find the equation of a circle in the $xy$ −plane with center $(-1,2)$ and a radius with an endpoint $(2,6)$.

     A. $(x - 1)^2 + (y + 2)^2 = 25$

     B. $(x + 1)^2 + (y - 2)^2 = 25$

     C. $(x - 1)^2 + (y + 2)^2 = \sqrt{5}$

     D. $(x + 1)^2 + (y - 2)^2 = 5$

17) What are the coordinates at the minimum point of $f(x) = x^2 + 2x - 1$?

    A. $(-1, 2)$

    B. $(1, -2)$

    C. $(-1, -2)$

    D. $(1, 2)$

18) What is the value of $x$ in this rational equation $\frac{2}{x+1} = \frac{x}{x^2-1}$?

    A. 1

    B. 2

    C. 3

    D. 4

19) Convert the radian measure $\frac{2\pi}{3}$ to degree measure.

    A. $120°$

    B. $60°$

    C. $90°$

    D. $150°$

20) The population of a colony increases according to equation $P = 1.7e^{rt}$, where $t$ is the number of months, and $r$ is the rate of growth. Which equation solves for $r$?

    A. $r = \dfrac{ln\left(\frac{1.7}{p}\right)}{t}$

    B. $r = \dfrac{t}{ln\left(\frac{1.7}{p}\right)}$

    C. $r = \dfrac{ln\left(\frac{p}{1.7}\right)}{t}$

    D. $r = \dfrac{t}{ln\left(\frac{p}{1.7}\right)}$

21) If $f(x) = 2x^3 + 4$ and $g(x) = \frac{1}{x}$, what is the value of $f(g(x))$?

   A. $\dfrac{1}{2x^3+4}$

   B. $\dfrac{2}{x^3} + 4$

   C. $2x^3 + 4 + \dfrac{1}{x}$

   D. $2x^2 + \dfrac{4}{x}$

22) In which direction does the graph of the parabola $y = -\frac{1}{2}(2 - x)^2 + 1$ open?

   A. Up

   B. Left

   C. Right

   D. Down

23) Convert 150 degrees to radian.

   A. $\dfrac{\pi}{6}$

   B. $\dfrac{\pi}{12}$

   C. $\dfrac{5\pi}{12}$

   D. $\dfrac{5\pi}{6}$

24) What is the domain and range of the radical function $y = 3\sqrt{2x + 4} + 8$?

   A. Domain $x \geq -2$, range $f(x) \geq 8$

   B. Domain $-\infty \leq x \leq +\infty$, range $-\infty \leq x \leq +\infty$

   C. Domain $x \geq -2$, range $-\infty \leq x \leq +\infty$

   D. Domain $x > -2$, range $f(x) \geq 8$

25) Solve this equation for $x$: $e^{2x} = 12$.

   A. $\frac{1}{2} ln(12)$

   B. $ln(12)$

   C. $ln(2)$

   D. $\frac{1}{2} ln(2)$

26) Simplify. $(-3 + 9i)(3 + 5i)$
   A. $54 + 12i$
   B. $-54 + 12i$
   C. $54 - 12i$
   D. $-54 - 12i$

27) A line passes through the point $(9,5)$ and has a slope of $-\frac{2}{3}$. Write an equation in point-slope form for this line.
   A. $y + 5 = \frac{2}{3}(x - 9)$
   B. $y - 5 = \frac{2}{3}(x + 9)$
   C. $y - 5 = -\frac{2}{3}(x - 9)$
   D. $y + 5 = \frac{-2}{3}(x - 9)$

28) Mr. Jackson has $1,000 in Citibank. He pays $500 to the bank every month. How much will he save after 2 years?
   A. $12,500
   B. $13,000
   C. $168,000
   D. $162,000

29) If $\theta$ is an acute angle and $sin\,\theta = \frac{4}{5}$, then $cos\,\theta =$?
   A. $\frac{3}{5}$
   B. $\frac{4}{3}$
   C. $\frac{3}{4}$
   D. $\frac{5}{3}$

30) Which best describes the graph of $\frac{x^2}{49} + \frac{y^2}{16} = 1$?

    A. Hyperbola

    B. Circle

    C. Parabola

    D. Elipse

31) If the center of a circle is at the point $(0, -3)$ and its circumference equals to $20\pi$, what is the equation of the circle's standard form?

    A. $(x - 3)^2 + (y)^2 = 2\pi$

    B. $(x + 3)^2 + (y)^2 = 10$

    C. $(x)^2 + (y + 3)^2 = 100$

    D. $(x)^2 + (y - 3)^2 = 20\pi$

32) If $tan\,\theta = \frac{5}{12}$ and $sin\,\theta > 0$, then $cos\,\theta =$?

    A. $\frac{12}{13}$

    B. $-\frac{5}{13}$

    C. $-\frac{12}{13}$

    D. $\frac{13}{12}$

33) What is the value of $y$ in the following system of equations?

$$2x + 5y = 11$$
$$4x - 2y = -14$$

    A. 3

    B. $-2$

    C. $-3$

    D. 2

34) If $f(x) = 2x^3 + 5x^2 + 2x$ and $g(x) = -3$, what is the value of $f(g(x))$?

   A. $-15$

   B. $2x^3 + 5x^2 + 2x + 3$

   C. $-3$

   D. $-6x^3 - 15x^2 - 6x$

35) What is the $y-$intercept of $y = x^2 + 2x - 3$?

   A. $(1,0)$

   B. $(-3,0)$

   C. $(0,-3)$

   D. $(0,1)$

36) If $sin\ A = \frac{1}{4}$ is a right triangle and angle $A$ is an acute angle, then what is $cos\ A$?

   A. $\frac{1}{4}$

   B. $\frac{\sqrt{15}}{4}$

   C. $\sqrt{15}$

   D. $-\frac{\sqrt{15}}{4}$

37) Solve the equation: $log_3(x + 20) - log_3(x + 2) = 1$.

   A. 7

   B. 22

   C. 3

   D. 14

38) Find a positive and a negative coterminal angle to angle $125°$.

   A. $55°$, $125°$

   B. $305°$, $-55°$

   C. $-235°$, $485°$

   D. $215°$, $35°$

39) If $\begin{bmatrix} -1 & -5 & -2 \\ 5 & 0 & x \end{bmatrix} \begin{bmatrix} y \\ 0 \\ 2 \end{bmatrix} = \begin{bmatrix} -8 \\ 28 \end{bmatrix}$, what is $x$ and $y$.

    A. $x = 4$ and $y = -4$

    B. $x = -4$ and $y = -1$

    C. $x = 4$ and $y = 4$

    D. $x = -1$ and $y = 4$

40) What is the domain and range of function: $f(x) = \frac{1}{x+2}$?

    A. $\{-2\}$ and $\mathbb{R}$

    B. $\mathbb{R}$ and $\mathbb{R} - \{0\}$

    C. $\mathbb{R} - \{2\}$ and $\mathbb{R} - \{0\}$

    D. $-\infty < x < +\infty$ and $\mathbb{R}$

41) What is the interval solution to the inequality $\frac{6x+12}{x-8} > 0$?

    A. $(-2, +\infty)$

    B. $(-\infty, 8) \cup (8, +\infty)$

    C. $(-\infty, 8) \cup (-2, +\infty)$

    D. $(-\infty, -2) \cup (8, +\infty)$

42) What is the parent graph of the following function and what transformations have taken place on it $y = (x + 2)^2$:

    A. The parent graph is $y = x^2$, which is shifted 2 units up.

    B. The parent graph is $y = x^2$, which is shifted 2 units down.

    C. The parent graph is $y = x^2$, which is shifted 2 units left.

    D. The parent graph is $y = x^2$, which is shifted 2 units right.

43) Perform the indicated operation and write the result in standard form:
$$(-7 + 12i)(-5 - 15i)$$
A. $215 + 45i$
B. $-215 + 45i$
C. $-215 - 45i$
D. $215 - 45i$

44) The profit (in dollars) from a car wash is given by the function $P(a)$, $P(a) = \frac{40a - 500}{a} + b$, where $a$ is the number of cars washed and $b$ is a constant. If 50 cars were washed today for a total profit of \$600, what is the value of $b$?
A. 500
B. 600
C. 570
D. 350

45) What is the sum of the first 7 terms of the series 1, 3, 9, ...?
A. 729
B. 2187
C. 1093
D. 364

46) What is the parent graph of the following function and what transformations have taken place on it?
$$y = 2 - (2 + 4x - x^2)$$
A. The parent graph is $y = -x^2$, which is shifted 4 units down and shifted 2 units left.
B. The parent graph is $y = x^2$, which is shifted 4 units up and shifted 2 units right.
C. The parent graph is $y = -x^2$, which is shifted 2 units right and shifted 4 units up.
D. The parent graph is $y = x^2$, which is shifted 2 units right and shifted 4 units down.

47) If $|A| = 2$, where $A$ is a matrix and represented by $\begin{bmatrix} 1 & 0 & 2 \\ 2 & -1 & q \\ 3 & -1 & 1 \end{bmatrix}$. Which is $q$?

    A. $-1$

    B. 0

    C. 1

    D. 2

48) Find the equation of the graphed hyperbola.

    A. $\dfrac{(x+3)^2}{4} - \dfrac{(y-4)^2}{9} = 1$

    B. $\dfrac{(x-3)^2}{4} - \dfrac{(y+4)^2}{9} = 1$

    C. $\dfrac{(y+4)^2}{4} - \dfrac{(x-3)^2}{9} = 1$

    D. $\dfrac{(y-4)^2}{4} - \dfrac{(x+3)^2}{9} = 1$

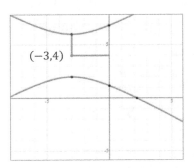

$(-3,4)$

49) If the cotangent of an angle $\beta$ is 1, then the tangent of angle $\beta$ is $\cdots$?

    A. 0

    B. $-1$

    C. $-\infty$

    D. 1

50) Write in terms of $log(r)$, $log(s)$, $log(t)$: $log \dfrac{s\sqrt{t}}{r^2}$.

    A. $\dfrac{log\, s \cdot \frac{1}{2} log\, t}{2\, log\, r}$

    B. $log\, s + 2\, log\, t - \dfrac{1}{2} log\, r$

    C. $log\, s + \dfrac{1}{2} log\, t - 2\, log\, r$

    D. $log\, s - \dfrac{1}{2} log\, t - 2\, log\, r$

51) What are the zeroes of the function $f(x) = x^3 + 6x^2 + 8x$?

A. $\{0, -2, -4\}$

B. $\{2, 4\}$

C. $\{-2, -4\}$

D. No roots

52) If $\frac{1}{2}log_{2x} 64 = 3$, what is the value of $x$?

A. $-2$

B. $4$

C. $1$

D. $3$

53) If $f(x) = 2x^3 + x$ and $g(x) = x - 2$, what is the value of $(f - g)(1)$?

A. $1$

B. $2$

C. $3$

D. $4$

54) What is the value of $log_3 \frac{\sqrt{27}}{9}$?

A. $3$

B. $0$

C. $1$

D. $-\frac{1}{2}$

55) Which statement describes the equation $x = y^2 + 2y - 4$?

A. It is a vertical parabola.

B. It is a vertical hyperbola.

C. It is a horizontal parabola.

D. It is a horizontal hyperbola.

56) What is the equation (in standard form) of the following graph?

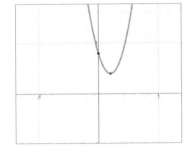

A. $y = 2x^2 - 4x - 4$
B. $y = 2(x - 1)^2 + 1$
C. $y = 2(x - 1)^2 - 1$
D. $y = 2x^2 - 4x + 4$

57) What is the inverse of $f(x) = 1 - x$?

A. $f^{-1}(x) = 1 - x$
B. $f^{-1}(x) = x - 1$
C. $f^{-1}(x) = \frac{1}{1-x}$
D. $f^{-1}(x) = \frac{1}{x-1}$

58) Which function is the invers of $f(x) = ln \sqrt{x}$?

A. $e^x$
B. $x^{\frac{1}{2}}$
C. $e^{2x}$
D. $\frac{1}{f(x)}$

59) What is the slope of a line that is perpendicular to the line $4x - 2y = 14$?

A. 2
B. $-2$
C. $\frac{1}{2}$
D. $-\frac{1}{2}$

60) What is the inverse of the function $f(x) = x^2 + 1$?

A. $f^{-1}(x) = \sqrt{x} - 1$
B. $f^{-1}(x) = \frac{1}{\sqrt{x-1}}$
C. $f^{-1}(x) = \frac{1}{x^2-1}$
D. $f^{-1}(x) = \pm\sqrt{x - 1}$

**End of Algebra 2 Practice Test 2.**

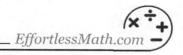

# Algebra 2 Practice Tests Answer Keys

Now, it's time to review your results to see where you went wrong and what areas you need to improve.

| Algebra 2 Practice Test 1 | | | | | | Algebra 2 Practice Test 2 | | | | |
|---|---|---|---|---|---|---|---|---|---|---|---|
| 1 | B | 21 | D | 41 | B | 1 | C | 21 | B | 41 | D |
| 2 | B | 22 | B | 42 | A | 2 | C | 22 | D | 42 | C |
| 3 | B | 23 | C | 43 | C | 3 | C | 23 | D | 43 | A |
| 4 | B | 24 | B | 44 | A | 4 | D | 24 | A | 44 | C |
| 5 | B | 25 | C | 45 | D | 5 | A | 25 | A | 45 | C |
| 6 | D | 26 | D | 46 | B | 6 | B | 26 | B | 46 | D |
| 7 | D | 27 | D | 47 | C | 7 | A | 27 | C | 47 | C |
| 8 | D | 28 | A | 48 | C | 8 | B | 28 | A | 48 | D |
| 9 | A | 29 | B | 49 | B | 9 | B | 29 | A | 49 | D |
| 10 | B | 30 | A | 50 | C | 10 | B | 30 | D | 50 | C |
| 11 | C | 31 | B | 51 | B | 11 | D | 31 | C | 51 | A |
| 12 | D | 32 | C | 52 | B | 12 | D | 32 | A | 52 | C |
| 13 | D | 33 | B | 53 | D | 13 | A | 33 | A | 53 | D |
| 14 | D | 34 | D | 54 | C | 14 | C | 34 | A | 54 | D |
| 15 | D | 35 | D | 55 | A | 15 | B | 35 | C | 55 | C |
| 16 | B | 36 | A | 56 | B | 16 | B | 36 | B | 56 | D |
| 17 | C | 37 | A | 57 | D | 17 | C | 37 | A | 57 | A |
| 18 | A | 38 | B | 58 | A | 18 | B | 38 | C | 58 | C |
| 19 | D | 39 | A | 59 | C | 19 | A | 39 | C | 59 | D |
| 20 | A | 40 | D | 60 | B | 20 | C | 40 | C | 60 | D |

# Algebra 2 Practice Tests Answers and Explanations

# Algebra 2 Practice Tests 1 Explanations

## 1) Choice B is correct

First, make sure that it's possible to multiply the two matrices. (The number of columns in the 1st one should be the same as the number of rows in the second one.) Therefore, choice B and C is impossible. Then, we have:

$[2 \quad 1] \times \begin{bmatrix} 0 & 2 \\ -1 & 0 \end{bmatrix} = [(2)(0) + (1)(-1) \quad (2)(2) + (1)(0)] = [-1 \quad 4]$, this is true.

$[2 \quad 0] \times \begin{bmatrix} 1 & 0 \\ 1 & 3 \end{bmatrix} = [(2)(1) + (0)(1) \quad (2)(0) + (0)(3)] = [2 \quad 0]$, this is NOT true!

## 2) Choice B is correct

To find any term in an arithmetic sequence, use this formula: $a_n = a_1 + d(n-1)$.

We have $a_3 = 13$ and $a_7 = 25$. Therefore: $a_3 = a_1 + d(3-1) \rightarrow a_1 + 2d = 13$ and $a_7 = a_1 + d(7-1) \rightarrow a_1 + 6d = 25$.

Solve the following equation:

$\begin{cases} a_1 + 2d = 13 \\ a_1 + 6d = 25 \end{cases}$

Subtracting the second equation from the first equation, we have:

$(a_1 + 6d) - (a_1 + 2d) = 25 - 13$. Simplify to: $a_1 + 6d - a_1 - 2d = 25 - 13 \rightarrow 4d = 12 \rightarrow d = 3$. By substituting $d = 3$ in the first equation, we get:

$a_1 + 2d = 13 \rightarrow a_1 + 2 \times 3 = 13 \rightarrow a_1 = 7$. Then, use this formula: $S_n = \frac{n}{2}(a_1 + a_n)$.

Then follow with: $S_n = \frac{n}{2}(a_1 + a_n) \rightarrow S_n = \frac{n}{2}(a_1 + a_1 + d(n-1)) \rightarrow S_n = \frac{n}{2}(2a_1 + d(n-1))$.

Therefore: $S_{10} = \frac{10}{2}(2(7) + 3(10-1)) \rightarrow S_{10} = 5(14 + 27) \rightarrow S_{10} = 205$.

### 3) Choice B is correct

Draw the graph corresponding to each function.
Rewrite:
$f(x) = 2x^2 - 4x \to f(x) = 2(x^2 - 2x) \to f(x) = 2(x-1)^2 - 2$.
According to a standard form of a parabola, $(1, -2)$
is the vertex. In addition, if $f(x) = 0$, then:
$2x^2 - 4x = 0 \to 2x(x-2) = 0 \to x = 0, x = 2$,

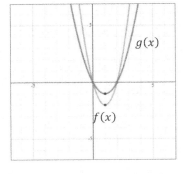

and $(0,0)$ and $(2,0)$ are $x-$intercepts for $f(x)$.
Correspondingly, we have for $g(x)$, $(1, -1)$ as its vertex and $(0,0)$ and $(2,0)$
are $x-$intercepts.
Therefore, they have the same $x-$intercepts.

### 4) Choice B is correct

First, factor of expression in the form of exponential of order 3: $64x^5 = 2^6 \times x^3 \times x^2$.

Then: $\sqrt[3]{64x^5} = \sqrt[3]{2^6 x^3 x^2} = 2^2 x \sqrt[3]{x^2}$.

Now use this rule: $\sqrt[n]{x^a} = x^{\frac{a}{n}}$. Therefore: $2^2 x \sqrt[3]{x^2} = 2^2 x x^{\frac{2}{3}} = 4x^{\frac{5}{3}}$.

### 5) Choice B is correct

If $a^y = x$, then: $\log_a x = y$. Therefore: $e^{\ln 4} = x \to \ln x = \ln 4$.

Since $\log_a b = \log_a c \to b = c$. Accordingly: $\ln x = \ln 4 \to x = 4$.

### 6) Choice D is correct

Use fractions division rule: $\frac{a}{b} \div \frac{c}{d} = \frac{a}{b} \times \frac{d}{c} = \frac{a \times d}{b \times c}$. Then: $\frac{5x}{x+3} \div \frac{x}{2x+6} = \frac{5x}{x+3} \times \frac{2x+6}{x} = \frac{5x(2x+6)}{x(x+3)} = \frac{5x \times 2(x+3)}{x(x+3)}$. Cancel common factor: $\frac{5x \times 2(x+3)}{x(x+3)} = \frac{10x(x+3)}{x(x+3)} = 10$.

### 7) Choice D is correct

Range: since $x^2$ is never negative, $x^2 + 2$ is never less than 2.

Hence, the range of $f(x)$ is "all real numbers where $f(x) \geq 2$".

### 8) Choice D is correct

Considering that $(fog)(x) = f(g(x))$, then:
$f(g(x)) = f(-2) = \frac{1}{2}(-2) - 1 = -2$.
Therfore, line $N$ corresponds to the answer.

### 9) Choice A is correct

We know $log_a \frac{x}{y} = log_a x - log_a y$. Therefore:

$x = 3\left(log_2 \frac{1}{32}\right) \rightarrow x = 3(log_2 1 - log_2 32)$. In addition, we know $log_a 1 = 0$.

Therefore: $x = -3 \, log_2 32$. Considering: $log_a x^b = b \times log_a x$, and $log_a a = 1$.

Then: $log_2 32 = log_2 2^5 = 5 \, log_2 2 = 5$. Finally, we have:

$x = -3 \, log_2 32 \rightarrow x = -3 \times 5 \rightarrow x = -15$.

### 10) Choice B is correct

We know that: $log_a b - log_a c = log_a \frac{b}{c}$ and $log_a b = c \rightarrow b = a^c$.

Then: $log_4(x + 2) - log_4(x - 2) = 1 \rightarrow log_4 \frac{x+2}{x-2} = 1 \rightarrow \frac{x+2}{x-2} = 4^1 = 4 \rightarrow x + 2 = 4(x - 2)$. Therefore: $x + 2 = 4x - 8 \rightarrow 4x - x = 8 + 2 \rightarrow 3x = 10 \rightarrow x = \frac{10}{3}$.

### 11) Choice C is correct

According to the trigonometric circle and $tan \, \theta = \frac{opposite}{adjacent}$, and since, $sin \, \theta < 0$ and $tan \, \theta > 0$, therefore, $cos \, \theta < 0$. Considering, $tan \, \theta = \frac{5}{12}$, then:

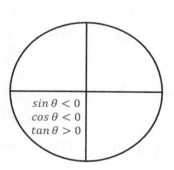

$$sin \, \theta < 0$$
$$cos \, \theta < 0$$
$$tan \, \theta > 0$$

$c = \sqrt{5^2 + 12^2} = \sqrt{25 + 144} = \sqrt{169} = 13,$

$cos \, \theta = \frac{adjacent}{hypotenuse} \rightarrow cos \, \theta = -\frac{12}{13} \, cos \, \theta = -\frac{12}{13}$.

### 12) Choice D is correct

The equation of a circle in standard form is: $(x - h)^2 + (y - k)^2 = r^2$.

Where the center is at: $(h, k)$ and its radius is $r$.

Then, for the circle with equation $(x - 4)^2 + (y + 7)^2 = 3$, the center is at $(4, -7)$ and its radius is $\sqrt{3}$ $(r^2 = 3 \rightarrow r = \sqrt{3})$.

### 13) Choice D is correct

Considering that $f(x)$ opens upward and $g(x)$ opens downward. Rewrite:
$f(x) = 2x^2 - x + 3 \rightarrow f(x) = 2\left(x - \frac{1}{4}\right)^2 + \frac{23}{8}$. So: $f(x) \geq \frac{23}{8}$.

On the other hand: $g(x) = -x^2 + 2x + 1 \rightarrow g(x) = -(x - 1)^2 + 2 \rightarrow g(x) \leq 2$.

Therefore, the minimum of $f(x)$ is greater than the maximum of $g(x)$.

## 14) Choice D is correct

To rationalize this imaginary expression, multiply both the numerator and denominator by the conjugate $\frac{12-i}{12-i}$. Then: $\frac{-8i}{12+i} = \frac{-8i(12-i)}{(12+i)(12-i)}$.

Apply complex arithmetic rule: $(a+bi)(a-bi) = a^2 + b^2$. Then:

$$\frac{-8i(12-i)}{(12+i)(12-i)} = \frac{-96i+8i^2}{12^2-i^2} = \frac{-96i-8}{144-(-1)} = \frac{-96i-8}{145} = -\frac{8}{145} - \frac{96}{145}i$$

## 15) Choice D is correct

To simplify this inequality, we need to isolate the variable on one side of the inequality. Add 6 to both sides: $-6 + 2x > -4 \rightarrow -6 + 2x + 6 > -4 + 6$.

Then: $2x > 2 \rightarrow \frac{2x}{2} > \frac{2}{2} \rightarrow x > 1$.

## 16) Choice B is correct

The function $f(x)$ is undefined when the denominator of $\frac{1}{(x-3)^2+4(x-3)+4}$ is equal to zero. The expression $(x-3)^2 + 4(x-3) + 4$ is a perfect square.

$(x-3)^2 + 4(x-3) + 4 = ((x-3)+2)^2$ can be rewritten as $(x-1)^2$.

The expression $(x-1)^2$ is equal to zero if $x = 1$.

Therefore, the value of $x$, for which $f(x)$ is undefined, is 1.

## 17) Choice C is correct

Use the matrix determinant, then:

$\left\| \begin{bmatrix} 2 & 6 \\ 0 & 5 \end{bmatrix} \right\| = (2)(5) - (6)(0) = 10 - 0 = 10$, this is NOT true!

$\left\| \begin{bmatrix} 2 & 6 & 3 \\ 0 & 1 & 1 \\ 4 & 7 & 4 \end{bmatrix} \right\| = 2((1)(4) - (1)(7)) - 6((0)(4) - (1)(4)) + 3((0)(7) - (1)(4)) =$
6, this is NOT true!

$\left\| \begin{bmatrix} 1 & -2 \\ -5 & 0 \end{bmatrix} \right\| = (1)(0) - (-2)(-5) = 0 - 10 = -10$, this is true.

$\left\| \begin{bmatrix} 1 & 3 & 0 \\ 0 & 5 & 1 \\ 1 & -1 & 0 \end{bmatrix} \right\| = 1((5)(0) - 1(-1)) - 3((0)(0) - 1(1)) + 0((0)(-1) - (5)(1)) =$
4, this is NOT true!

### 18) Choice A is correct

According to the contents of the question, $a_1 = 15$ and $a_n = 135$.

Use this formula: $a_n = a_1 + (n - 1)d$. The answer is obtained by solving this equation: $135 = 15 + (n - 1)5$. Simplify, $120 = (n - 1)5 \rightarrow n - 1 = 24 \rightarrow n = 25$.

### 19) Choice D is correct

According to the graph, the function is a parabola that open downward and $(0, -2)$ is the vertex. Now, we have:

$$(y - k) = 4p(x - h)^2 \rightarrow (y - (-2)) = 4p(x - 0)^2 \rightarrow y = 4px^2 - 2, \text{ where } p < 0.$$

Since all points of the graph must be satisfed in the equation. Therefore, for point $(2, -4)$, we have:

$$y = 4px^2 - 2 \rightarrow -4 = 4p(2)^2 - 2 \rightarrow 16p = -2 \rightarrow p = -\frac{1}{8}.$$

Finally: $y = 4\left(-\frac{1}{8}\right)x^2 - 2 \rightarrow y = -\frac{1}{2}x^2 - 2.$

### 20) Choice A is correct

The standard form of a parabola is: $(y - k) = 4p(x - h)^2$. The coordinate $(-1, -5)$ is the vertex. Now, we have: $(y + 5) = 4p(x + 1)^2$. Put $(0, -3)$ in the equation. Then:

$$(-3 + 5) = 4p(0 + 1)^2 \rightarrow 2 = 4p \rightarrow p = \frac{1}{2}.$$

Finally: $(y + 5) = 4p(x + 1)^2 \rightarrow (y + 5) = 4\frac{1}{2}(x + 1)^2 \rightarrow (y + 5) = 2(x + 1)^2.$

Then simplify to:

$$(y + 5) = 2(x + 1)^2 \rightarrow y + 5 = 2(x^2 + 2x + 1) \rightarrow y = 2x^2 + 4x - 3.$$

### 21) Choice D is correct

To draw the graph of $y \leq -2x - 2$, you first need to graph the line: $y = -2x - 2$.
Since there is a less than ($\leq$) sign, draw a solid line.
The slope is $-2$ and the $y$ −intercept is $-2$.
Then, choose a testing point and substitute the value of $x$ and $y$ from that point into the inequality. The easiest point to test is the origin: $(0,0)$,
$(0,0) \rightarrow y < -2x - 2 \rightarrow 0 < -2(0) - 2 \rightarrow 0 < -2.$
This is incorrect! $0$ is not less than $-2$.
So, the left side of the line is the solution of this inequality.

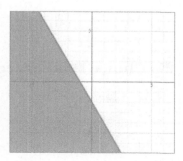

## 22) Choice B is correct

The standard form of an ellipse is: $\frac{(x-h)^2}{a^2} + \frac{(y-k)^2}{b^2} = 1$. Since center is at $(0,0)$,

rewrite this to: $\frac{(x-0)^2}{a^2} + \frac{(y-0)^2}{b^2} = 1 \rightarrow \frac{x^2}{a^2} + \frac{y^2}{b^2} = 1$.

When major axis has length of 9 units along the $x$ −axis, it means:

$x = \frac{9}{2}$, and also $b = 8$. Then: $\frac{x^2}{\left(\frac{9}{2}\right)^2} + \frac{y^2}{(8)^2} = 1 \rightarrow \frac{x^2}{\frac{81}{4}} + \frac{y^2}{64} = 1 \rightarrow \frac{4x^2}{81} + \frac{y^2}{64} = 1$.

## 23) Choice C is correct

The standard form of a up/down parabola is: $(y - k) = 4p(x - h)^2$.

Rewrite to standard form:

$y = (x - 2)^2 + 3 \rightarrow y - 3 = (x - 2)^2 \rightarrow (y - 3) = 4 \cdot \frac{1}{4}(x - 2)^2$.

Therefore, the vertex is $(2,3)$.

## 24) Choice B is correct

The equation of a circle in standard form is: $(x - h)^2 + (y - k)^2 = r^2$, where $r$ is the radius of the circle and $(-1,2)$ is the center. Considering that perimeter of the circle is $4\pi$, then: $r = 2$. Therefore: $(x + 1)^2 + (y - 2)^2 = 2^2 \rightarrow (x + 1)^2 + (y - 2)^2 = 4$.

## 25) Choice C is correct

Simplify and use this rule: $\frac{x^a}{x^b} = x^{a-b}$. Now: $\frac{81x^5}{3x^2} = 27x^{5-2} = 27x^2$. Then:

$\sqrt[3]{\frac{81x^5}{3x^2}} = \sqrt[3]{27x^3}$. Finally, using $\sqrt[n]{x^a} = x^{\frac{a}{n}}$, we have: $\sqrt[3]{27x^3} = (27x^3)^{\frac{1}{3}} =$

$(3^3x^3)^{\frac{1}{3}} = 3^{3 \times \frac{1}{3}} x^{3 \times \frac{1}{3}} = 3x$.

## 26) Choice D is correct

Since $(fog)(x) = f(g(x))$, then:

$f(-x - 2) = (-x - 2) + 1 = -x - 1$, this is NOT true!

$f(x - 2) = -(x - 2) + 1 = -x + 2 + 1 = -x + 3$, this is NOT true!

$f(x + 2) = (x + 2) + 1 = x + 3$, this is NOT true!

$f(-x + 2) = -(-x + 2) + 1 = x - 2 + 1 = x - 1$, this is true!

## 27) Choice D is correct

Use the logarithm rule that if $f(x) = g(x)$, then: $\log_a f(x) = \log_a g(x)$.

Therefore: $3(7^{2x}) = 11 \rightarrow \log 3(7^{2x}) = \log 11$.

Next: $\log_a(x \cdot y) = \log_a x + \log_a y \rightarrow \log 3(7^{2x}) = \log 3 + \log 7^{2x}$, and $\log_a x^b = b \log_a x \rightarrow \log 7^{2x} = 2x \log 7$. So: $3(7^{2x}) = 11 \rightarrow \log 3 + 2x \log 7 = \log 11$.

Rewrite this as: $2x \log 7 = \log 11 - \log 3 \rightarrow x = \frac{\log 11 - \log 3}{2 \log 7}$.

### 28) Choice A is correct

By definition, the $x$ −intercept is the point where a graph crosses the $x$ −axis. if $y = 0$ is put into the equation, it would become:

$y = \frac{x-1}{1-x^2} \rightarrow \frac{x-1}{1-x^2} = 0 \rightarrow x - 1 = 0 \rightarrow x = 1.$

However, we know that the function is not defined for $x = 1$ and $x = -1$, as these values would result is division by zero.

Therefore, the graph of $y = \frac{x-1}{1-x^2}$ would not have $x$ −intercepts.

### 29) Choice B is correct

Use the geometric sequence formula: $a_n = a_1 r^{n-1}$.

Now, substitute $a_1 = 2$ and $a_3 = \frac{1}{8}$. So, $a_3 = 2r^{3-1} = \frac{1}{8}$.

Therefore:

$r^2 = \frac{1}{16} \rightarrow r = \frac{1}{4}$ or $r = -\frac{1}{4}$. Then, $x = a_2 \rightarrow x = a_1 r^{2-1} \rightarrow x = 2r.$

Solve with one of value $r = \frac{1}{4}$ or $r = -\frac{1}{4}$. Hence, $x = \frac{1}{2}$ or $x = -\frac{1}{2}$.

### 30) Choice A is correct

To find $AC$, use sine $B$. Then: $\sin \theta = \frac{opposite}{hypotenuse}$.

$\sin 40° = \frac{AC}{6} \rightarrow 6 \times \sin 40° = AC.$

Now use a calculator to find: $\sin 40°$. $\sin 40° \approx 0.643 \rightarrow AC = 3.858 \approx 3.86.$

### 31) Choice B is correct

We know that: $i = \sqrt{-1} \rightarrow i^2 = -1$. Use FOIL (First-Out-In-Last) method:

$(-4 + 9i)(3 + 5i) = -12 - 20i + 27i + 45i^2 = -12 + 7i - 45 = -57 + 7i.$

### 32) Choice C is correct

Since this inequality contains absolute value, then, the value inside absolute value bars is greater than $-4$ and less than $4$. Therefore:

$|x - 10| \le 4 \rightarrow -4 \le x - 10 \le 4 \rightarrow -4 + 10 \le x - 10 + 10 \le 4 + 10.$

Then, $6 \le x \le 14.$

### 33) Choice B is correct

We know that: $\sin^2 a + \cos^2 a = 1$. Which means that:

$x + \sin^2 a + \cos^2 a = 3 \rightarrow x + 1 = 3 \rightarrow x = 2.$

### 34) Choice D is correct

We know that the graph $y = f(x - k)$; $k > 0$, is shifted $k$ units to the right of the graph $y = f(x)$ and $k < 0$, is shifted $k$ units to the left.

In addition, the function $y = f(x)$ is symmetric to the function $y = f(-x)$, with respect to the $y$−axis and symmetric to the function $y = -f(x)$ and to the $x$−axis.

Now, $y = log(-x)$ is symmetric to $y = log\,x$ with respect to the $y$−axis as follows:

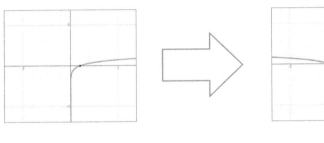

$y = log\,x$                     $y = log(-x)$

On the other hand, $y = log(1 - x)$ is shifted 1 unit to the right, since $y = log -(x - 1)$. Finally, $y = -log(1 - x)$ is symmetric with respect to the $x$−axis as follows:

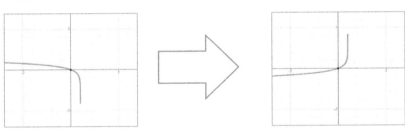

$y = log(1 - x)$                     $y = -log(1 - x)$

### 35) Choice D is correct

Coterminal angles are equal angles. To find a coterminal of an angle, add or subtract 360 degrees (or $2\pi$ for radian) to the given angle. Then:

Positive angle: $\frac{\pi}{2} + 2\pi = \frac{5\pi}{2}$.

Negative angle: $\frac{\pi}{2} - 2\pi = -\frac{3\pi}{2}$.

### 36) Choice A is correct

The standard form of the circle equation is: $(x - h)^2 + (y - k)^2 = r^2$ where the radius of the circle is $r$, and it's centered at $(h, k)$. First, move the loose number to the right side:
$x^2 + y^2 - 6x + 4y = -4$. Group $x$ −variables and $y$ −variables together:
$(x^2 - 6x) + (y^2 + 4y) = -4$. Convert $x$ to square form:
$(x^2 - 6x + 9) + (y^2 + 4y) = -4 + 9 \rightarrow (x - 3)^2 + (y^2 + 4y) = 5$.
Convert $y$ to square form:
$(x - 3)^2 + (y^2 + 4y + 4) = 5 + 4 \rightarrow (x - 3)^2 + (y + 2)^2 = 9$.
Then, the equation of the circle in standard form is: $(x - 3)^2 + (y + 2)^2 = 3^2$.
The center of the circle is at $(3, -2)$ and its radius is 3.

### 37) Choice A is correct

First, $y = -x$, then, replace all $x$'s with $y$ and all $y$'s with $x$: $x = -y$. Now, solve for $y$:
$x = -y \rightarrow -x = y$. Actually, the inverse of $y = -x$ is itself. In othr words, the invers of a graph is asymmetric to $y = x$.

### 38) Choice B is correct

The logarithm is another way of writing exponent. $log_b y = x$ is equivalent to $y = b^x$.
Rewrite the logarithm in exponent form: $log_2 x = 5 \rightarrow 2^5 = x \rightarrow x = 32$.

### 39) Choice A is correct

In a rational function, if the denominator has a bigger degree than the numerator, the horizontal asymptote is the $x$ −axis or the line $y = 0$.
In the function $f(x) = \frac{x+3}{x^2+1}$, the degree of the numerator is 1 ($x$ to the power of 1) and the degree of the denominator is 2 ($x$ to the power of 2). The horizontal asymptote is the line $y = 0$.

### 40) Choice D is correct

Here we can substitute 8 for $x$ in the equation. Thus, $y - 3 = 2(8 + 5)$, $y - 3 = 26$. Adding 3 to both sides of the equation: $y = 26 + 3 \rightarrow y = 29$.

### 41) Choice B is correct

The answer is as follows:
Plug in the values of $x$ and $y$ for each point in the inequality and check the result.
$(-1,2) \rightarrow y < 5x^2 - 2x - 5 \rightarrow 2 < 5(-1)^2 - 2(-1) - 5 \rightarrow 2 < 2$, this is NOT true!
$(3,6) \rightarrow y < 5x^2 - 2x - 5 \rightarrow 6 < 5(3)^2 - 2(3) - 5 \rightarrow 6 < 34$, this is true!
$(1,12) \rightarrow y < 5x^2 - 2x - 5 \rightarrow 12 < 5(1)^2 - 2(1) - 5 \rightarrow 12 < -2$, this is NOT true!
$(-1,6) \rightarrow y < 5x^2 - 2x - 5 \rightarrow 6 < 5(-1)^2 - 2(-1) - 5 \rightarrow 6 < 2$, this is NOT true!

### 42) Choice A is correct

Put the value of $S$ in the equation. $F = 9 \log \frac{10^{-5}}{10^{-6}}$. Simplify:

$F = 9 \log \frac{10^{-5}}{10^{-6}} \rightarrow F = 9 \log 10$. Using algorithm rule: $\log_a a = 1 \rightarrow \log 10 = 1$.

Therefore: $F = 9 \log 10 \rightarrow F = 9$.

### 43) Choice C is correct

For two function $f(x)$ and $g(x)$, we know that $(f - g)(x) = f(x) - g(x)$.
Therefore:
$$(g - f)(x) = g(x) - f(x) \rightarrow (-x^2 - 2x - 1) - (x + 6) = -x^2 - 2x - 1 - x - 6$$
$$= -x^2 - 3x - 7.$$

### 44) Choice A is correct

According to the graph, the function is quadratic and opens downward. Therefore, the parent function is equal to $x^2$. The parent function of the quadratic functions is $x^2$.

### 45) Choice D is correct

The value of $cos\, 30° = \frac{\sqrt{3}}{2}$.

### 46) Choice B is correct

The equation of a circle in standard form is: $(x - h)^2 + (y - k)^2 = r^2$, where $r$ is the radius of the circle. In this circle, the radius is 5:

$r^2 = 25 \rightarrow r = 5$, $(x + 2)^2 + (y - 4)^2 = 25$.

Area of a circle: $A = \pi r^2 = \pi(5)^2 = 25\pi$.

### 47) Choice C is correct

The function $f(x) = ln(x)$ with domain $x > 0$ and range $-\infty < f(x) < +\infty$ has an $x$ −intercept with coordinates $(1,0)$ where $x = 0$ is a vertical asymptote.

Its graph is as follows:

On the other hand, we know that the graph $y = f(x - k); k > 0$, is shifted $k$ units to the right of the graph $y = f(x)$ and if $k < 0$, is shifted $k$ units to the left.

Now, in the example graph, the $x$ −intercept with coordinates $(2,0)$ and vertical asymptote with equation $x = 1$ is shifted 1 unit to the right. Therefore, the function $y = ln(x - 1)$ is represented by an example graph.

**48) Choice C is correct**

Add the elements in the matching positions:

$$\begin{bmatrix} 2 & 4 \\ -5 & -1 \\ -2 & -6 \end{bmatrix} + \begin{bmatrix} a & 0 \\ 0 & 7 \\ 3 & 5 \end{bmatrix} = \begin{bmatrix} 2+a & 4 \\ -5 & 6 \\ 1 & -1 \end{bmatrix}.$$

Now, we have: $\begin{bmatrix} 2+a & 4 \\ -5 & 6 \\ 1 & -1 \end{bmatrix} = \begin{bmatrix} 1 & 4 \\ -5 & 6 \\ 1 & b \end{bmatrix}.$

This means that: $2 + a = 1 \rightarrow a = -1$ and $b = -1$.

**49) Choice B is correct**

The standard form of an up/down parabola is: $(y - k) = 4p(x - h)^2$.

According to the graph, the vertex is equal to $(-2, 4)$, and since the graph opens downward, $p < 0$.

On the other hand, $(0,0)$ also applies to the function. Then: $(y - k) = 4p(x - h)^2 \rightarrow (y - 4) = 4p(x + 2)^2$. Therefore: $(0 - 4) = 4p(0 + 2)^2 \rightarrow -4 = 4p(4) \rightarrow 16p = -4 \rightarrow p = -\frac{1}{4}$. Finally:

$$(y - 4) = 4\left(-\frac{1}{4}\right)(x + 2)^2 \rightarrow y - 4 = -x^2 - 4x - 4 \rightarrow y = -x^2 - 4x.$$

**50) Choice C is correct**

To find the value of $x$, use the cosine on the angle $x$:

$$\cos \theta = \frac{adjacent}{hypotenuse} \rightarrow \cos x = \frac{10}{14} = \frac{5}{7}$$

Use a calculator to find inverse cosine:

$$\cos^{-1}\left(\frac{5}{7}\right) = 44.42° \approx 44°$$

Which means that: $x = 44$.

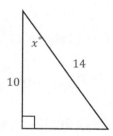

**51) Choice B is correct**

The sum of two supplementary angles is 180 degrees. Then: $(2x - 6) + (x + 3) = 180$.
Simplify and solve for $x$:
$(2x - 6) + (x + 3) = 180 \rightarrow 3x - 3 = 180 \rightarrow 3x = 183 \rightarrow x = 61$.

### 52) Choice B is correct

To solve for the side $AC$, we need to use the sine of angle $B$. Then:

$$sin\,\theta = \frac{opposite}{hypotenuse}, \; sin\,50° = \frac{AC}{6} \rightarrow 6 \times sin\,50° = AC.$$

Now use a calculator to find sine $50°$. $sin\,50° \approx 0.766$.

$AC = 6 \times 0.766 = 4.596$, rounding to the nearest tenth: $4.596 \approx 4.6$.

### 53) Choice D is correct

The answer is on the following graph:

First, simplify the inequality:

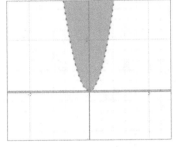

$$2y > 4x^2 \rightarrow \frac{2y}{2} > \frac{4x^2}{2} \rightarrow y > 2x^2.$$

Now, graph the quadratic $y = 2x^2$.

Plug in some values for $x$ and solve for $y$.

$x = 0 \rightarrow y = 2(0)^2 = 0$,

$x = 1 \rightarrow y = 2(1)^2 = 2$,

$x = -1 \rightarrow y = 2(-1)^2 = 2$,

$x = 2 \rightarrow y = 2(2)^2 = 8$,

$x = -2 \rightarrow y = 2(-2)^2 = 8$,

Since the inequality is greater than $(>)$, we need to use dash lines.

Now, choose a testing point inside the parabola. Let's choose $(0, 2)$.

$y > 2x^2 \rightarrow 2 > 2(0)^2 \rightarrow 2 > 0$.

This is true. Therefore, the inside of the parabola is the solution section.

### 54) Choice C is correct

Since $tan\,\theta = \frac{opposite}{adjacent}$, we have the following: $tan\frac{2\pi}{3} = \frac{sin\frac{2\pi}{3}}{cos\frac{2\pi}{3}} = \frac{\frac{\sqrt{3}}{2}}{-\frac{1}{2}} = -\sqrt{3}$.

### 55) Choice A is correct

To solve this system of equation, add the two equations. Then:

$$\begin{cases} x + 2y = 10 \\ 6x - 2y = 18 \end{cases} \rightarrow x + 6x + 2y - 2y = 10 + 18 \rightarrow 7x = 28 \rightarrow x = 4.$$

Substitute the value of $x$ in the first equation and solve for $y$:

$x + 2y = 10,\; x = 4 \rightarrow 4 + 2y = 10 \rightarrow 2y = 10 - 4 \rightarrow 2y = 6 \rightarrow y = 3.$

## 56) Choice B is correct

Solve the system of equations by elimination method.

$\begin{array}{l} 3x - 4y = -20 \\ \underline{-x + 2y = 20} \end{array}$. Multiply the second equation by 3, then add it to the first

equation.

$\begin{array}{l} 3x - 4y = -20 \\ \underline{3(-x + 2y = 20)} \end{array} \rightarrow \begin{array}{l} 3x - 4y = -20 \\ \underline{-3x + 6y = 60} \end{array}$. Now, adding the equations: $2y = 40 \rightarrow y =$

20.

## 57) Choice D is correct

Since this inequality contains an absolute value, then, the value inside absolute value bars is greater than 4 and less than $-4$. Therefore:

$|x - 2| \geq 4 \rightarrow 4 \leq x - 2$ or $x - 2 \leq -4 \rightarrow 4 + 2 \leq x - 2 + 2$ or

$x - 2 + 2 \leq -4 + 2$.

Then, $6 \leq x \cup x \leq -2$.

## 58) Choice A is correct

The answer is on the following graph:

A quadratic function in the vertex form is: $y = a(x - h)^2 + k$ with $(h, k)$ as the vertex. The vertex of $y = (x + 1)^2 - 2$ is $(-1, -2)$.

Substitute zero for $x$ and solve for $y$:

$y = (0 + 1)^2 - 2 = -1$.

The $y-$intercept is $(0, -1)$.

Now, you can simply graph the quadratic function. Notice that the quadratic function is a $U-$shaped curve. (You can plug in values of $x$ and solve for $y$ to get some points on the graph.)

## 59) Choice C is correct

Find the factors of the denominators: $\frac{3x}{x^2-36} = \frac{3x}{(x-6)(x+6)}$ and $\frac{4}{2x-12} = \frac{4}{2(x-6)}$.

Since the factor $(x - 6)$ is common in both denominators, then, the least common denominator is: $2(x - 6)(x + 6) = 2x^2 - 72$.

**60) Choice B is correct**

We know that: $sin^2 a + cos^2 a = 1$.

Therefore: $3x - 2 sin^2 a - 2 cos^2 a = 4 \rightarrow 3x - 2(sin^2 a + cos^2 a) = 4 \rightarrow 3x - 2 = 4 \rightarrow 3x = 4 + 2 = 6 \rightarrow x = 2$.

# Algebra 2 Practice Tests 2 Explanations

**1) Choice C is correct**

Rewrite: $R = \frac{(5000x - x^2)}{200} \rightarrow R = -\frac{1}{200}(x^2 - 5000x)$.

Then: $R = -\frac{1}{200}\left(x - \frac{5000}{2}\right)^2 + 31250$.

Therefore, this equation is a vertical parabola that opens downward.

The maximum amount of the equation is equal to vertex. According to the equation, $(2500, 31250)$ is the vertex. That means that the maiximum amount is 31,250.

**2) Choice C is correct**

First, find the radius of the circle. The circumference of a circle $= 2\pi$.

Circumference $= 2\pi r \rightarrow r = 1$. The equation of a circle in the coordinate plane:

$(x - h)^2 + (y - k)^2 = r^2$. Center: $(h, k)$ and radius: $r$. The center is at the point $(-4, 2)$, so: $(-4, 2) \rightarrow h = -4, k = 2$. Then, the equation of the circle is:

$(x + 4)^2 + (y - 2)^2 = 1$.

**3) Choice C is correct**

We know that: $\tan \theta = \frac{opposite}{adjacent}$, and $\tan \theta = \frac{8}{15}$, therefore, the opposite side of

the angle $x$ is 8 and the adjacent side is 15. Let's draw the

triangle. Using the Pythagorean theorem, we have:

$a^2 + b^2 = c^2 \rightarrow 8^2 + 15^2 = c^2 \rightarrow 64 + 225 = c^2 \rightarrow c = 17$,

$\sin x = \frac{opposite}{hypotenuse} = \frac{8}{17}$.

**4) Choice D is correct**

To simplify the fraction, multiply both the numerator and denominator by $i$.

$\frac{4 - 3i}{-4i} \times \frac{i}{i} = \frac{4i - 3i^2}{-4i^2}$, $i^2 = -1$. Then: $\frac{4i - 3i^2}{-4i^2} = \frac{4i - 3(-1)}{-4(-1)} = \frac{4i + 3}{4} = \frac{4i}{4} + \frac{3}{4} = \frac{3}{4} + i$.

**5) Choice A is correct**

Based on corresponding members from two matrices, we get:

$\begin{cases} 2x = x + 3y - 5 \\ 4x = 2y + 10 \end{cases} \rightarrow \begin{cases} x - 3y = -5 \\ 4x - 2y = 10 \end{cases}$.

Multiply the first equation by $-4$. $\begin{cases} -4x + 12y = 20 \\ 4x - 2y = 10 \end{cases}$.

Then add two equations: $10y = 30 \rightarrow y = 3 \rightarrow x = 4 \rightarrow x \times y = 12$.

**6) Choice B is correct**

To find the $\left(\frac{f}{g}\right)(x)$, divide $f(x)$ by $g(x)$. $f(x) = -8x^2 + 12x$ and $g(x) = -2x +$

3. Then: $\left(\frac{f}{g}\right)(x) = \frac{f(x)}{g(x)} = \frac{-8x^2 + 12x}{-2x + 3}$.

Factor the numerator $(-8x^2 + 12x)$. Take the common factor $4x$ out:

$-8x^2 + 12x = 4x(-2x + 3) \rightarrow \frac{-8x^2 + 12x}{-2x + 3} = \frac{4x(-2x + 3)}{-2x + 3}$.

Cancel out the common factor: $(-2x + 3)$: $\frac{4x(-2x + 3)}{-2x + 3} = 4x$.

**7) Choice A is correct**

Use the slope equation: $m = \frac{y_2 - y_1}{x_2 - x_1} \rightarrow \frac{7 - 3}{-9 - (-4)} = \frac{4}{-9 + 4} = \frac{4}{-5} \rightarrow m = -\frac{4}{5}$.

The slope of the line is $-\frac{4}{5}$.

**8) Choice B is correct**

Rewrite $P = 120N - 0.2N^2$ as a standard form of a parabola:

$P = 120N - 0.2N^2 \rightarrow P = -0.2(N^2 - 600N) \rightarrow P = -0.2(N - 300)^2 + 18000$.

As the equation is a downward facing parabola, then the vertex is the maximum value. Therefore, $(300, 18000)$ is the vertex.

The maximum profit will be for the company to rent out 300 cars.

**9) Choice B is correct**

Consider the relationship among all sides of the special right triangle $30°$-$60°$-$90°$ is provided in this triangle:

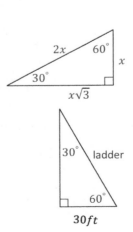

In this triangle, the opposite side of the $30°$ angle is half of the hypotenuse. Draw the shape of this question: The ladder is the hypotenuse.

$\cos 60° = \frac{adjacent}{hypotenuse} \rightarrow \frac{1}{2} = \frac{30}{ladder} \rightarrow ladder = 60$.

Therefore, the ladder is $60ft$.

## 10) Choice B is correct

Use the cross multiplication to solve for $x$:

$\frac{5}{x+1} = \frac{x+1}{x^2-1} \rightarrow 5(x^2 - 1) = (x + 1)(x + 1)$.

Simplify $5(x^2 - 1)$ using the distributive property:

$5(x^2 - 1) = 5x^2 - 5$.

Simplify $(x + 1)(x + 1)$ using the FOIL (First-Out-In-Last) method:

$(x + 1)(x + 1) = x^2 + x + x + 1 = x^2 + 2x + 1$.

Then: $5(x^2 - 1) = (x + 1)(x + 1) \rightarrow 5x^2 - 5 = x^2 + 2x + 1$.

Since this is a quadratic equation, we need to bring all terms to one side of the equation. Subtract $(x^2 + 2x + 1)$ from both sides.

Then: $5x^2 - 5 - (x^2 + 2x + 1) = x^2 + 2x + 1 - (x^2 + 2x + 1)$.

Simplify and combine like terms:

$5x^2 - 5 - x^2 - 2x - 1 = x^2 + 2x + 1 - (x^2 + 2x + 1)$.

Then: $4x^2 - 2x - 6 = 0$.

Use the quadratic formula to solve for $x$: $x_{1,2} = \frac{-b \pm \sqrt{b^2 - 4ac}}{2a}$.

Then: $x_{1,2} = \frac{-b \pm \sqrt{b^2 - 4ac}}{2a} = \frac{-(-2) \pm \sqrt{(-2)^2 - 4(4)(-6)}}{2(4)} = \frac{2 \pm \sqrt{100}}{8}$.

Therefore:

$x = \frac{2+10}{8} = \frac{12}{8} = \frac{3}{2}$, or $x = \frac{2-10}{8} = \frac{-8}{8} = -1$.

The solution $x = -1$ is not defined in the original equation. (If $x = -1$, then the denominators are equal to zero, which is not defined.) Therefore, the solution $x = \frac{3}{2}$ is the only acceptable solution.

## 11) Choice D is correct

When the logarithms have the same base: $f(x) = g(x)$, then: $x = y$,

$log(5x + 2) = log(3x - 1) \rightarrow (5x + 2) = (3x - 1) \rightarrow 5x + 2 - 3x + 1 = 0$

$\rightarrow 2x + 3 = 0 \rightarrow 2x = -3 \rightarrow x = -\frac{3}{2}$

Verify Solution: $log\left(5\left(-\frac{3}{2}\right) + 2\right) = log\left(\frac{-15}{2} + 2\right) = log\left(-\frac{11}{2}\right) = log(-5.5)$.

Logarithms of negative numbers are not defined. Therefore, there is no solution for this equation.

## 12) Choice D is correct

Since this inequality contains an absolute value, the value inside the absolute value bars is greater than $-4$ and less than $4$. Then:

$x - 2 \geq 4 \rightarrow x \geq 4 + 2 \rightarrow x \geq 6$, or $x - 2 \leq -4 \rightarrow x \leq -4 + 2 \rightarrow x \leq -2$.

The solution is: $x \leq -2 \cup x \geq 6$.

## 13) Choice A is correct

Using this rule: $(xy)^a = x^a \times y^a$, then: $(x^3 y^5)^{\frac{2}{3}} = (x^3)^{\frac{2}{3}}(y^5)^{\frac{2}{3}}$.

On the other hand, going by this rule: $(x^a)^b = x^{a \times b}$, we have:

$(x^3)^{\frac{2}{3}}(y^5)^{\frac{2}{3}} = x^{3 \times \frac{2}{3}} y^{5 \times \frac{2}{3}} = x^2 y^{\frac{10}{3}}$. Factoring and using this rule: $\sqrt[n]{x^a} = x^{\frac{a}{n}}$, we

have: $x^2 y^{\frac{10}{3}} = x^2 y^{\frac{9}{3}} \times y^{\frac{1}{3}} = x^2 y^3 \times y^{\frac{1}{3}} = x^2 y^3 \sqrt[3]{y}$.

## 14) Choice C is correct

To find the vertical asymptote(s) of a rational function, set the denominator to 0 and solve for $x$. Then: $3x + 4 = 0 \rightarrow 3x = -4 \rightarrow x = -\frac{4}{3}$.

The vertical asymptote is $x = -\frac{4}{3}$.

## 15) Choice B is correct

The first term is 4 and the common ratio is 3, therefore, $a_1 = 4$ and $r = 3$.

The finite geometric series formula is:

$S_n = \sum_{i=1}^{n} ar^{i-1} = a_1 \left(\frac{1-r^n}{1-r}\right)$. So, $484 = 4\left(\frac{1-3^n}{1-3}\right)$.

Solve the obtained equation:

$484 = 4\left(\frac{1-3^n}{1-3}\right) \rightarrow 484 = (-2)(1 - 3^n) \rightarrow 3^n = 243 \rightarrow 3^n = 3^5 \rightarrow n = 5$.

## 16) Choice B is correct

The equation of a circle can be written as:

$(x - h)^2 + (y - k)^2 = r^2$, where $(h, k)$ is the coordinates of the center of the circle and $r$ is the radius of the circle.

Since the coordinates of the center of the circle are $(-1, 2)$, the equation is:

$(x + 1)^2 + (y - 2)^2 = r^2$, where $r$ is the radius. The radius of the circle is the distance from the center $(-1, 2)$, to the given endpoint of a radius, $(2, 6)$. By the distance formula,

$r^2 = (2 - (-1))^2 + (6 - 2)^2 = (3)^2 + (4)^2 = 9 + 16 = 25$.

Therefore, the equation of the given circle is: $(x + 1)^2 + (y - 2)^2 = 25$.

**17) Choice C is correct**

Rewrite: $f(x) = x^2 + 2x - 1 \rightarrow f(x) = (x^2 + 2x + 1) - 2 \rightarrow f(x) = (x + 1)^2 - 2$.

This means that $(-1, -2)$ is the minimum point.

**18) Choice B is correct**

Use the cross multiply method if $\frac{a}{b} = \frac{c}{d}$, then: $a \times d = b \times c$.

$\frac{2}{x+1} = \frac{x}{x^2-1} \rightarrow 2(x^2 - 1) = x(x + 1)$. Then: $2x^2 - 2 = x^2 + x$.

Now, substract both sides by $(x^2 + x)$.

Then: $2x^2 - 2 - (x^2 + x) = x^2 + x - (x^2 + x) \rightarrow x^2 - x - 2 = 0 \rightarrow$
$(x + 1)(x - 2) = 0$.

Finally: $x = -1$ and $x = 2$, where $x = -1$ is the root of the denominator and is undefined. Therefore, $x = 2$.

**19) Choice A is correct**

Use this formula: $Degrees = Radian \times \frac{180}{\pi}$,

$$Degrees = \frac{2\pi}{3} \times \frac{180}{\pi} = \frac{360\pi}{3\pi} = 120$$

**20) Choice C is correct**

If $f(x) = g(x)$, then: $\log_a f(x) = \log_a g(x)$.

So, $\ln P = \ln 1.7 e^{rt}$. Use the logarithm rule: $\log_a(x \cdot y) = \log_a x + \log_a y$.

Therefore, we have: $\ln P = \ln 1.7 + \ln e^{rt}$.

Then: $\log_a x^y = y \log_a x$ and $\log_a a = 1$.

Accordingly: $\ln P = \ln 1.7 + rt \cdot \ln e \rightarrow \ln P = \ln 1.7 + rt$.

Rewrite as: $rt = \ln P - \ln 1.7$.

$\log_a x - \log_a y = \log_a \frac{x}{y} \rightarrow rt = \ln\left(\frac{P}{1.7}\right)$.

Then divide both sides by $t$:

$$r = \frac{\ln\left(\frac{P}{1.7}\right)}{t}$$

**21) Choice B is correct**

To find $f(g(x))$, substitute $x$ with $\frac{1}{x}$ in the function $f(x)$. Then:

$$f(g(x)) = f\left(\frac{1}{x}\right) = 2 \times \left(\frac{1}{x}\right)^3 + 4 = \frac{2}{x^3} + 4.$$

**22) Choice D is correct**

Rewrite to:

$$y = -\frac{1}{2}(2-x)^2 + 1 \rightarrow y - 1 = -\frac{1}{2}(2-x)^2 \rightarrow y - 1 = 4\left(-\frac{1}{8}\right)(2-x)^2.$$

So: $y - 1 = 4\left(-\frac{1}{8}\right)(x-2)^2$. The standard form of a up/down parabola is:

$$(y - k) = 4p(x - h)^2.$$

Accordingly, $(2,1)$ is the vertex and $p < 0$, the graph opens downward.

**23) Choice D is correct**

Use this formula: $Radian = Degrees \times \frac{\pi}{180}$.

$$Radian = 150 \times \frac{\pi}{180} = \frac{150\pi}{180} = \frac{5\pi}{6}$$

**24) Choice A is correct**

For domain: Find non-negative values for radicals: $2x + 4 \geq 0$.
the domain of functions: $2x + 4 \geq 0 \rightarrow 2x \geq -4 \rightarrow x \geq -2$.
Domain of the function $y = 3\sqrt{2x + 4} + 8$: $x \geq -2$.
For the range: The range of a radical function in the form $c\sqrt{ax + b} + k$ is:
$f(x) \geq k$.
For the function $y = 3\sqrt{2x + 4} + 8$, the value of $k$ is 8. Then: $f(x) \geq 8$.
Range of the function $y = 3\sqrt{2x + 4} + 8$: $f(x) \geq 8$.

**25) Choice A is correct**

If $f(x) = g(x)$, then: $ln(f(x)) = ln(g(x)) \rightarrow ln(e^{2x}) = ln(12)$.
Use the logarithm rule:
$log_a x^b = b \, log_a x \rightarrow ln(e^{2x}) = 2x \, ln(e) \rightarrow (2x) \, ln(e) = ln(12)$.
Since: $ln(e) = 1$, then: $(2x) \, ln(e) = ln(12) \rightarrow 2x = ln(12) \rightarrow x = \frac{ln(12)}{2}$.

**26) Choice B is correct**

Use the FOIL (First, Out, In, Last) method to multiply two imaginary expressions:

$$(-3 + 9i)(3 + 5i) = -9 - 15i + 27i + 45i^2.$$

We know that:

$i^2 = -1$. Then: $-9 - 15i + 27i + 45i^2 = -9 + 12i - 45 = -54 + 12i$.

### 27) Choice C is correct

The "point-slope" form of the equation of a straight line is: $y - y_1 = m(x - x_1)$.

The slope of the line is $-\frac{2}{3}$ and the point provided is: $(9, 5)$.

Then: $y - y_1 = m(x - x_1)$, where $m = -\frac{2}{3}$, $(x_1, y_1) = (9, 5)$.

Therefore: $y - 5 = -\frac{2}{3}(x - 9)$.

The equation in the point-slope form of the line is: $y - 5 = -\frac{2}{3}(x - 9)$.

### 28) Choice A is correct

Use arithmetic sequence beginning with 1000 and common difference 500 for the 24th month:

$$a_n = a_1 + (n - 1)d \rightarrow a_{24} = 1000 + (24 - 1)500 \rightarrow a_{24} = 12500.$$

### 29) Choice A is correct

Since: $sin\,\theta = \frac{opposite}{hypotenuse} = \frac{4}{5}$. Now, we have the following triangle. Then:

$c = \sqrt{5^2 - 4^2} = \sqrt{25 - 16} = \sqrt{9} = \pm 3$, so:

$cos\,\theta = \frac{adjacent}{hypotenuse} = \frac{3}{5}$ or $cos\,\theta = -\frac{3}{5}$.

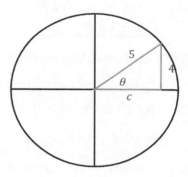

We know that $\theta$ is an acute angle. Then, $cos\,\theta$ is positive. Therefore, $cos\,\theta = \frac{3}{5}$.

### 30) Choice D is correct

Considering the classification of conic sections, the equation is a standard form of horizontal ellipse:

$\frac{(x-h)^2}{a^2} + \frac{(y-k)^2}{b^2} = 1$, where $a > b$.

### 31) Choice C is correct

First, find the radius of the circle. The circumference of a circle $= 20\pi$.

Circumference $= 2\pi r = 20\pi \rightarrow r = 10$.

The equation of a circle in the coordinate plane:

$(x - h)^2 + (y - k)^2 = r^2$.

Center: $(h, k)$ and radius: $r$. The center of the circle is at the point $(0, -3)$:

$(0, -3) = (h, k) \to h = 0, k = -3$.

Then, the equation of the circle is: $(x)^2 + (y + 3)^2 = 100$.

### 32) Choice A is correct

We know that: $\tan \theta = \frac{opposite}{adjacent}$, $\tan \theta = \frac{5}{12}$. So, we have the following right triangle. Then:

$c = \sqrt{5^2 + 12^2} = \sqrt{25 + 144} = \sqrt{169} = 13$,

$\cos \theta = \frac{adjacent}{hypotenuse} = \frac{12}{13}$.

### 33) Choice A is correct

Solving systems of equations by elimination:

Multiply the first equation by $(-2)$, then add it to the second equation.

$\begin{matrix} -2(2x + 5y = 11) \\ 4x - 2y = -14 \end{matrix} \to \begin{matrix} -4x - 10y = -22 \\ 4x - 2y = -14 \end{matrix} \to -12y = -36 \to y = 3$.

### 34) Choice A is correct

$f(g(x))$ for two function $f(x)$ and $g(x)$, means that the output from $g(x)$ becomes the input for $f(x)$. We have: $g(x) = -3$.

Then: $f(g(x)) = f(-3) = 2(-3)^3 + 5(-3)^2 + 2(-3) = -54 + 45 - 6 = -15$.

### 35) Choice C is correct

The $y$−intercept is a point corresponding to $x = 0$, then: $y = (0)^2 + 2(0) - 3 \to y = -3$. Therefore, $(0, -3)$ is the $y$−intercept.

### 36) Choice B is correct

$\theta$ is an acute angle. Then, the trigonometric ratios for $\theta$ are positive. $\sin A = \frac{1}{4}$. Since $\sin \theta = \frac{opposite}{hypotenuse}$, we have the following right triangle. Then:

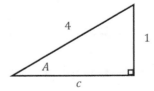

$c = \sqrt{4^2 - 1^2} = \sqrt{16 - 1} = \sqrt{15}$.

$\cos A = \frac{adjacent}{hypotenuse} = \frac{\sqrt{15}}{4}$.

### 37) Choice A is correct

$log_3(x + 20) - log_3(x + 2) = 1$. First, condense the two logarithms:

$log_3(x + 20) - log_3(x + 2) = 1 \rightarrow log_3\left(\frac{x+20}{x+2}\right) = 1$.

We know that: $log_a a = 1$. Then:

$log_3\left(\frac{x+20}{x+2}\right) = 1 \rightarrow log_3\left(\frac{x+20}{x+2}\right) = log_3 3 \rightarrow \frac{x+20}{x+2} = 3$.

Use cross multiplication and solve for $x$.

$\frac{x+20}{x+2} = 3 \rightarrow x + 20 = 3(x + 2) \rightarrow x + 20 = 3x + 6 \rightarrow 2x = 14 \rightarrow x = 7$.

### 38) Choice C is correct

To find the coterminal angle to angle 125°:
$125° + 360° = 485°$

$125° - 360° = -235°$

### 39) Choice C is correct

Multiply the rows of the first matrix by the columns of the second matrix:

$\begin{bmatrix} -1 & -5 & -2 \\ 5 & 0 & x \end{bmatrix}\begin{bmatrix} y \\ 0 \\ 2 \end{bmatrix} = \begin{bmatrix} (-1)(y) + (-5)(0) + (-2)(2) \\ (5)(y) + (0)(0) + (x)(2) \end{bmatrix} = \begin{bmatrix} -y - 4 \\ 5y + 2x \end{bmatrix} = \begin{bmatrix} -8 \\ 28 \end{bmatrix}$.

Then: $-y - 4 = -8 \rightarrow -y = -4 \rightarrow y = 4$.

Therefore, substitute 4 with $y$ in: $5y + 2x = 28$.

Finally: $5(4) + 2x = 28 \rightarrow 20 + 2x = 28 \rightarrow 2x = 8 \rightarrow x = 4$.

### 40) Choice C is correct

The function is not defined for $x = -2$, as this value would result is division by zero.

Hence the domain of $f(x)$ is all real numbers except $-2$.

Range: No matter how large or small $x$ becomes, $f(x)$ will never be equal to zero.

Therefore, the range of $f(x)$ is all real numbers except zero.

**41) Choice D is correct**

Factor the numerator: $\frac{6x+12}{x-8} > 0 \rightarrow \frac{6(x+2)}{x-8} > 0$.

Divide both sides by 6: $\frac{\frac{6(x+2)}{x-8}}{6} > \frac{0}{6} \rightarrow \frac{x+2}{x-8} > 0$.

Determine the signs of the factors $\frac{x+2}{x-8}$.

Find the roots of the numerator and denominator:

$x + 2 = 0 \rightarrow x = -2$ and $x - 8 = 0 \rightarrow x = 8$.

Plug in some values of $x$ of the intervals: $(-\infty, 2]$, $[2,8)$ and $(8, +\infty)$. Now, check the solutions.

Only $x < -2$, or $x > 8$ work in the inequality.

The interval notation: $(-\infty, -2) \cup (8, +\infty)$.

**42) Choice C is correct**

We know that the graph $y = f(x - k); k > 0$, shifted $k$ units to the right of the graph $y = f(x)$, and $k < 0$ shifted $k$ units to the left. Then, $y = (x + 2)^2$, is shifted 2 units left of the graph $y = x^2$.

**43) Choice A is correct**

Use the FOIL (First, Out, In, Last) method to multiply two imaginary expressions:

$(-7 + 12i)(-5 - 15i) =$

$(-7)(-5) + (-7)(-15i) + (12i)(-5) + (12i)(-15i) = 35 + 105i - 60i - 180i^2$.

Combine like terms $(+105i - 60i)$ and simplify:

$35 + 105i - 60i - 180i^2 = 35 + 45i - 180i^2$, $i^2 = -1$.

Then: $35 + 45i - 180i^2 = 35 + 45i - 180(-1) = 35 + 45i + 180 = 215 + 45i$.

**44) Choice C is correct**

This is a simple matter of substituting values for variables. We are given that the 50 cars were washed today, therefore we can substitute that for $a$. Giving us the expression $P(50) = \frac{40(50)-500}{50} + b$. We are also given that the profit was $600, which we can substitute for $P(a)$. The equation is then $600 = \frac{40(50)-500}{50} + b$.

Simplifying the fraction gives us the equation $600 = 30 + b$.

Subtracting both sides of the equation by 30 gives us $b = 570$, which is the answer.

### 45) Choice C is correct

Using the finite geometric series formula:

$S_n = \sum_{i=1}^{n} ar^{i-1} = a_1 \left( \frac{1-r^n}{1-r} \right)$.

Substitute $a_1 = 1$ and $r = 3$ in the previous formula.

$S_7 = 1 \left( \frac{1-3^7}{1-3} \right) = \frac{1-2187}{1-3} = \frac{-2186}{-2} \rightarrow S_7 = 1093$.

### 46) Choice D is correct

Rewrite: $y = 2 - (2 + 4x - x^2) \rightarrow y = x^2 - 4x \rightarrow y = (x-2)^2 - 4$.

This means that the parent graph is $y = x^2$, which is shifted 2 units right and shifted 4 units down.

### 47) Choice C is correct

First, use the matrix determinant for $\begin{bmatrix} a & b & c \\ d & e & f \\ g & h & i \end{bmatrix}$ as follows:

$|A| = a(ei - fh) - b(di - fg) + c(dh - eg)$.

Then: $|A| = 1\big((-1)(1) - (q)(-1)\big) - 0\big((2)(1) - (q)(3)\big) + 2\big((2)(-1) - (-1)(3)\big) = -1 + q + 2 = q + 1$.

Since $|A| = 2$, then: $q + 1 = 2 \rightarrow q = 1$.

### 48) Choice D is correct

According to the graph, we have the standard form of an up/down hyperbola with center $(-3,4)$, semi-axis 2 and semi-conjugate-axis 3. Then:

$\frac{(y-k)^2}{a^2} - \frac{(x-h)^2}{b^2} = 1 \rightarrow \frac{(y-4)^2}{4} - \frac{(x+3)^2}{9} = 1$.

### 49) Choice D is correct

The cotangent is the reciprocal of tangent: $\tan \beta = \frac{1}{\cot \beta} = \frac{1}{1} = 1$.

### 50) Choice C is correct

Use the logarithms rule: $\log_a x - \log_a y = \log_a \frac{x}{y}$. Then: $\log \frac{s\sqrt{t}}{r^2} = \log s\sqrt{t} - \log r^2$.

Using the logarithms rule: $\log_a (x \cdot y) = \log_a x + \log_a y$, we get:

$\log s\sqrt{t} = \log s + \log \sqrt{t}$. We know that $\sqrt{t} = t^{\frac{1}{2}}$. Then:

$\log s + \log \sqrt{t} = \log s + \log t^{\frac{1}{2}}$.

Now, use the logarithms rule: $\log_a x^n = n \log_a x$. Then:

$\log t^{\frac{1}{2}} = \frac{1}{2} \log t$ and $\log r^2 = 2 \log r$.

Finally: $\log \frac{s\sqrt{t}}{r^2} = \log s\sqrt{t} - \log r^2 = \log s + \frac{1}{2} \log t - 2 \log r$.

**51) Choice A is correct**

First, factor the function: $f(x) = x^3 + 6x^2 + 8x = x(x + 4)(x + 2)$.

To find the zeros, $f(x)$ should be zero. $f(x) = x(x + 4)(x + 2) = 0$.

Therefore, the zeros are: $x = 0$, $(x + 4) = 0 \rightarrow x = -4$, $(x + 2) = 0 \rightarrow x = -2$.

**52) Choice C is correct**

Use the logarithm rule: $\log_a x^b = b \log_x x \rightarrow \frac{1}{2} \log_{2x} 64 = \log_{2x} 64^{\frac{1}{2}}$.

In addition: $\log_a b = c \rightarrow a^c = b$.

Now, we have: $\frac{1}{2} \log_{2x} 64 = 3 \rightarrow \log_{2x} 64^{\frac{1}{2}} = 3 \rightarrow \log_{2x} 8 = 3 \rightarrow (2x)^3 = 8$.

Finally, simplify: $8x^3 = 8 \rightarrow x^3 = 1 \rightarrow x = 1$.

**53) Choice D is correct**

Since $(f - g)(x) = f(x) - g(x)$, then:

$(f - g)(x) = (2x^3 + x) - (x - 2) = 2x^3 + 2$.

Therefore, replace $x$ with 1:

$(f - g)(1) = 2(1)^3 + 2 = 4$.

**54) Choice D is correct**

Simplify $\log_3 \frac{\sqrt{27}}{9} = \log_3 \frac{3^{\frac{3}{2}}}{3^2} = \log_3 3^{\frac{3}{2}-2} = \log_3 3^{-\frac{1}{2}}$.

We know that: $\log_a b = c$ is equivalent to $b = a^c$.

Then: $\log_3 3^{-\frac{1}{2}} = -\frac{1}{2} \log_3 3 = -\frac{1}{2}$.

**55) Choice C is correct**

Rewrite as: $x = (y^2 + 2y + 1 - 1) - 4 \rightarrow x = (y + 1)^2 - 5 \rightarrow (x + 5) = (y + 1)^2$.

The equation $(x + 5) = 4 \cdot \left(\frac{1}{4}\right)(y + 1)^2$ is like the standard form of a left/right parabola, $(x - h) = 4p(y - k)^2$. Which means that it is a horizontal parabola.

## 56) Choice D is correct

To figure out what the equation of the graph is, first find the vertex. From the graph we can determine that the vertex is at (1,2).

We can use vertex form to solve the equation of this graph. Recall vertex form, $y = a(x - h)^2 + k$, where $h$ is the $x$ −coordinate of the vertex, and $k$ is the $y$ −coordinate of the vertex.

Plugging in our values, you get $y = a(x - 1)^2 + 2$.

To solve for $a$, we need to pick a point on the graph and plug it into the equation. Such as $(0,4)$, $4 = a(0 - 1)^2 + 2$. Then: $4 = a(-1)^2 + 2 \rightarrow 4 = a + 2 \rightarrow a = 2$. Now the equation is $y = 2(x - 1)^2 + 2$.

Then expand to:

$y = 2(x^2 - 2x + 1) + 2 \rightarrow y = 2x^2 - 4x + 2 + 2 \rightarrow y = 2x^2 - 4x + 4$.

## 57) Choice A is correct

Replace $f(x)$ with $y$: $y = 1 - x$. Then, replace all $x$'s with $y$ and all $y$'s with $x$: $x = 1 - y$. Now, solve for $y$:

$x = 1 - y \rightarrow x - 1 = -y \rightarrow -y = x - 1 \rightarrow y = 1 - x$.

Finally, replace $y$ with $f^{-1}(x)$: $f^{-1}(x) = 1 - x$.

## 58) Choice C is correct

Replace $f(x)$ with $y$: $y = \ln \sqrt{x}$. Then, replace all $x$'s with $y$ and all $y$'s with $x$. Write as: $x = \ln \sqrt{y} \rightarrow x = \ln y^{\frac{1}{2}}$. We know: $\log_a x^b = b \log_a x$.

Then: $x = \ln y^{\frac{1}{2}} \rightarrow x = \frac{1}{2} \ln y$. Multiply both sides with 2: $2x = \ln y$.

Using, $y = \log_a x \rightarrow x = a^y$. Which means that $e^{2x} = y$.

Finally: $y = e^{2x} \rightarrow f^{-1}(x) = e^{2x}$.

## 59) Choice D is correct

The equation of a line in slope intercept form is: $y = mx + b$.

Solve for $y$: $4x - 2y = 14 \rightarrow -2y = 14 - 4x \rightarrow y = \frac{(14 - 4x)}{-2} \rightarrow y = 2x - 7$.

The slope of the line is 2. The slope of the line perpendicular to this line is:

$m_1 \times m_2 = -1 \rightarrow 2 \times m_2 = -1 \rightarrow m_2 = -\frac{1}{2}$.

**60) Choice D is correct**

First, replace $f(x)$ with $y$: $y = x^2 + 1$.

Then, replace all $x$'s with $y$ and all $y$'s with $x$: $x = y^2 + 1$.

Now, solve for $y$: $x = y^2 + 1 \rightarrow x - 1 = y^2 \rightarrow |y| = \sqrt{x - 1} \rightarrow y = \pm\sqrt{x - 1}$.

Finally, replace $y$ with $f^{-1}(x)$: $f^{-1}(x) = \pm\sqrt{x - 1}$.